Transversal Enterprises in the Drama of Shakespeare and his Contemporaries

Also by Bryan Reynolds

PERFORMING TRANSVERSALLY: Reimagining Shakespeare and the Critical Future

BECOMING CRIMINAL: Transversal Performance and Cultural Dissidence in Early Modern England

REMATERIALIZING SHAKESPEARE: Authority and Representation on the Early Modern English Stage *(co-editor with William N. West)*

SHAKESPEARE WITHOUT CLASS: Misappropriations of Cultural Capital *(co-editor with Donald Hedrick)*

Transversal Enterprises in the Drama of Shakespeare and his Contemporaries

Fugitive Explorations

Bryan Reynolds

First published 2006 by
PALGRAVE MACMILLAN
Houndmills, Basingstoke, Hampshire RG21 6XS and
175 Fifth Avenue, New York, N.Y. 10010
Companies and representatives throughout the world

PALGRAVE MACMILLAN is the global academic imprint of the Palgrave Macmillan division of St. Martin's Press, LLC and of Palgrave Macmillan Ltd. Macmillan® is a registered trademark in the United States, United Kingdom and other countries. Palgrave is a registered trademark in the European Union and other countries.

ISBN 13: 978–1–4039–3211–2   hardback
ISBN 10: 1–4039–3211–5   hardback

This book is printed on paper suitable for recycling and made from fully managed and sustained forest sources.

A catalogue record for this book is available from the British Library.

Library of Congress Cataloging-in-Publication Data

Reynolds, Bryan (Bryan Randolph)
    Transversal enterprises in the drama of Shakespeare and his contemporaries: fugitive explorations / by Bryan Reynolds.
        p. cm.
    Includes bibliographical references and index.
    ISBN 1–4039–3211–5 (cloth)
        1. English drama–Early modern and Elizabethan, 1500–1600–History and criticism. 2. Subjectivity in literature. 3. Shakespeare, William, 1564–1616–Criticism and interpretation. 4. Popular culture–England–History–16th century. 5. Shakespeare, William, 1564–1616–Contemporaries. I. Title.

PR658.S79R49 2006
822'.309353–dc22   2005052719

10   9   8   7   6   5   4   3   2   1
15   14   13   12   11   10   09   08   07   06

Transferred to digital printing 2006

For my daughter Sky,
Who sunshines my every day,
Who wide eyes my future,
And gazes back with me smiling.

Contents

List of Figures ix

Acknowledgments x

1 Transversal Poetics and Fugitive Explorations: 1
Theaterspace, Paused Consciousness, Subjunctivity,
and *Macbeth*
Bryan Reynolds

2 The Reckoning of Moll Cutpurse: Transversal 27
Reimaginings of *The Roaring Girl*
Bryan Reynolds & Janna Segal

3 The Delusion of Critique: Subjunctive Space, 64
Transversality, and the Conceit of Deceit in *Hamlet*
Anthony Kubiak & Bryan Reynolds

4 Comedic Law: Projective Transversality, Deceit 85
Conceits, and the Conjuring of *Macbeth* and
Doctor Faustus in Jonson's *The Devil is an Ass*
Amy Cook & Bryan Reynolds

5 I Might Like You Better If We Slept Together: 112
the Historical Drift of Place in *The Changeling*
Donald Hedrick & Bryan Reynolds

6 Fugitive Explorations in *Romeo and Juliet*: Searching for 124
Transversality inside the Goldmine of R&Jspace
Bryan Reynolds & Janna Segal

7 Viewing Antitheatricality: or, *Tamburlaine*'s Post-Theater 168
Bryan Reynolds & Ayanna Thompson

8 Becomings Roman/Comings-to-be Villain: Pressurized 183
Belongings and the Coding of Ethnicity, Religion, and
Nationality in Peele & Shakespeare's *Titus Andronicus*
Glenn Odom & Bryan Reynolds

9 Awakening the Werewolf Within: Self-help, Vanishing 227
Mediation, and Transversality in *The Duchess of Malfi*
Courtney Lehmann & Bryan Reynolds

10 Performative Transversations: Collaborations Through   240
    and Beyond Greene's *Friar Bacon and Friar Bungay*
    Bryan Reynolds & Henry Turner

Afterword: Re: connaissance   251
Bruce Smith

A Very Short List of Writings Significant to the Development   257
of Transversal Poetics

Notes on the Collaborators   260

Index   263

List of Figures

8.1 "Titus Andronicus" (drawing, Henry Peacham, 1595,   214
     reproduced by permission of the Marquess of Bath,
     Longleat House)
8.2 "The lamentable and tragicall history of   215
     Titus Andronicus" (woodcut, 1658, reproduced
     by permission of the Folger Shakespeare Library)

Acknowledgments

When working on *Performing Transversally: Reimagining Shakespeare and the Critical Future* (Palgrave Macmillan, 2003), my collaborators and I got so enthusiastic about researching and writing together that the topic of a second book repeatedly came up. To be sure, the specter of this book powerfully preceded its own manifestation: its affective presence, fantastically occupying our imaginations, irresistibly inspired us to go for it. To what end, well, that was going to be a surprise. As the project got underway, the scope became more and more expansive and challenging. The parameters and interventions, stakes and aspirations, concepts and emotions were all explored, negotiated, and heightened. Although occasionally frantic and delirious, we became increasingly zealous about the project as the ideas became more enthralling and addictive, as difficulties surfaced and were remedied, as gateways into new territories opened and we ventured through them. *Transversal Enterprises in the Drama of Shakespeare and his Contemporaries: Fugitive Explorations* is the product of a team.

The two years that it took to write this book was a magnificent adventure. It was also during this time that William West and I co-edited *Rematerializing Shakespeare: Authority and Representation on the Early Modern English Stage* (Palgrave Macmillan, 2005), and that I co-founded the Transversal Theater Company, which – in addition to performances in the United States – took productions of two plays of mine, *Unbuckled* and *Woof, Daddy*, on tour in Romania (2004) and Poland (2005), respectively. Most importantly, my wife Kris and I collaborated on bringing our daughter Sky into the world on August 19, 2003 and on conceiving our son Zephyr, born July 1, 2005. Never before have I done so many things with so many wonderful people with such speed and intensity, and never before have I felt more creative and exhilarated. For this great gift, I am indebted to many people, some of whom I am delighted to thank here.

This book could only have been written with the help of my collaborators, so it is to them that I am supremely grateful. Thank you. I am grateful for the generous support given to me by the University of California at Irvine and by its Claire Trevor School of the Arts and Department of Drama in particular. Among the friends, family, and many colleagues at Irvine and beyond who I want to dearly thank for supporting me in different ways during the writing of this book are: Nohema Fernandez, Douglas-Scott Goheen, Eli Simon, Cipriana Petre, Alicia Tycer, Allison Crilly, Daphne Lei, Cliff Faulkner, Stacey Haggin, Richard Brestoff, Lonnie Alcaraz, Madeline Kozlowski, Marianne McDonald, Stephen Barker, Jane Newman, Viviana Comensoli, Péter Dávidházi, Jim Carmody, Marie Rutkoski, Jason Cruz, Felice Weis, Barbara Thibodeau, Janelle Reinelt, Robert Cohen, Cam Harvey, Bridget Matz, Dennis Lang, Joan Rasp, Gil Harris, Jason Hillyard, Michel Lang, Martine van Elk, Nick Radel, Arthur Kinney, David Razin, Lisa Razin, Shelly Razin, Janet Razin, Gary Curtis, Nancy Curtis, Dan Donoghue, Carolyn Odom, Linda Charnes, Craig Dionne, Joe Porter, Steve Mentz, Jim Harner, Paddy Fumerton, Bill Engel, Ann Pellegrini, Bruce Boehrer, Tony Dawson, Simon Palfrey, Jeffrey Kahan, Tim Murray, Michele Musacchio, Chris Hughes, Peter Eversmann, Elaine Aston, Indira Ghose, Ewan Fernie, Brian Burgoon, Hendriekje de Jong, Robert Mateescu, Rodney Byrd, Todd Wright, Greg Ungar, Richard Burt, Greg Reynolds, Donna Reynolds, Don Reynolds, Marge Garber, Ian Munro, Becky Helfer, Will West, Matt Grant, Anke Ortlepp, Kirsten Lammich, James Intriligator, Margo Crespin, Jennifer Gjulameti, Jenn Colella, Tom Augst, Scott Albertson, Patrick Williams, Karen Weber, and Kim Savelson.

I am especially grateful to my editor, Paula Kennedy, and to her assistant, Helen Craine, as well as to the rest of Palgrave Macmillan's excellent staff, for shepherding this book through all the necessary stages with remarkable efficiency.

Beyond words, I am indebted to my lovely wife Kris.

Portions of this book have appeared in *Rogues and Early Modern English Culture*, edited by Craig Dionne and Steve Mentz (Ann Arbor: University of Michigan Press, 2004) and *Early Theatre*, Volume 7:2 (2004); and are here reprinted by kind permission.

1

Transversal Poetics and Fugitive Explorations: Theaterspace, Paused Consciousness, Subjunctivity, and *Macbeth*

Bryan Reynolds

Transversal relations

The 2003–2004 war on Iraq, launched in the wake of the 2001 attacks on the United States and the subsequent invasion of Afghanistan, emphasized for me the need for activism, critical inquiry, and pedagogy that is rigorous, theoretical, and socially and politically engaged. Societies worldwide are rapidly becoming more interconnected through mass media, transportation, and commerce, and the means by which people compete and negotiate for resources and power are becoming progressively multifaceted, industrious, insightful, and desperate. Capitalist governments and corporations demonstrate repeatedly that a combination of ingenuity and deception – for example, the Bush and Blair administrations' ability to equivocate about such things as evidence for weapons of mass destruction – is requisite for the successful manipulation and dominance of discourses, societies, markets, and nations.

Developments in technology, culture, and science give the impression that societies are moving faster and getting "better," and previously uncharted terrains – ideational and/or material – are being explored, graphed, occupied, or subsumed. Exponentially, super-powered nations (the United States, Great Britain, China) and super-empowered iconic concepts and institutions (Christianity, Capitalism, Islamic Jihad) generate via "sociopolitical conductors" influential products, including texts, artworks, systems, and students. "Transversal poetics" is the evolving critical approach, theory, and aesthetics that I have developed in collaboration with others to foster agency, creativity, and the production of more conscientious and socially purposeful scholarship and pedagogy.[1] Sociopolitical conductors are vital to transversal poetics because

1

they are the familial, religious, juridical, media, and educational structures – the replicators, transmitters, and orchestrators of thoughts, meanings, and desires – that interconnect a society's ideological and cultural framework.[2] The interrelations among the conductors and their products generate conceptually dynamic assemblages, or "articulatory spaces," which are discursive environments that surround, enmesh, embody, and laminate charged topics, objects, and events. These emergent formations enhance and further the conductors' interactions with and disseminations of "open power," "state power," and/or "transversal power."[3]

Open power is any power that does not fall under the categories of state power or transversal power. State power is any force that works in the interest of coherence and organization among any variables; for instance, state power is at work as you read this and impose order and meaning to facilitate comprehension. Transversal power is any force – physical or ideational, friendly or antagonistic – that inspires emotional, conceptual, and/or material deviations from the established norms for any variables, whether individuated or forming a group. In agreement with their investment in social, cultural, economic, and political determinations, sociopolitical conductors work across spacetime, consciously and/or not, either to convert open and/or transversal powers into state power or, less typically, to unleash transversal and/or state power. In both cases, sociopolitical conductors discursively influence the articulatory spaces fueled by and fueling the respective powers, which escalate and radiate both diachronically and synchronically within and through articulatory spaces. Thus, social and cultural economies negotiate and function, vis-à-vis sociopolitical conductors, in conjunction with the articulatory spaces through which they develop. Through this engagement people come to see and believe certain things, consequently undergoing "becomings" and "comings-to-be" as the result of exposures and performances.

Becomings are desiring processes by which people transform into something different – physically, conceptually, and/or emotionally – from what they were, and if they were identified and normalized by a dominant force, such as state law, religious credo, cultural aesthetic, or official language, then any change in them is becomings-other. Alternatively, comings-to-be occur when people lose control during the process of becomings-other and become more of/or something else than anticipated or preferred. In other words, becomings are active processes, often self-inaugurated and pursued intentionally, whereas comings-to-be, however induced by becomings, are generated by the energies, ideas, people, societies, and so on to which the subject aspires, is drawn, or encounters by happenstance. For instance, if someone wants to become a member of a surfer subculture in California, she would need to actively and self-consciously become like other members in that community by adopting the dress, speech, and behavioral codes used by the community's members to distinguish themselves from others and readily identify each other. Such assimilation normally requires observation, imitation and emulation, and therefore the infusion of certain units of information, cultural substances, or, perhaps, what zoologist Richard Dawkins, in *The Selfish Gene*, calls "memes" (192).[4]

Dawkins's neologism is an amalgam of the words "memory," "imitation," and "gene," and he gives as examples of memes, "tunes, ideas, catch-phrases, clothes fashions, ways of making pots or building arches" (192). According to Dawkins, "Just as genes propagate themselves in the gene pool by leaping from body to body via sperms or eggs, so memes propagate themselves in the meme pool by leaping from brain to brain via a process which, in the broad sense, can be called imitation" (192). He explains that "imitation" refers to copy as in genetic "replication," but as a result of cultural rather than natural selection: "If a scientist hears, or reads about, a good idea, he passes it onto his colleagues and students. ... If the idea catches on, it can be said to propagate itself, spreading from brain to brain" (192). As a concrete example, Dawkins asks his readers to consider "the idea of God":

> How does it replicate itself? By the spoken and written word, aided by great music and great art. Why does it have such high survival value? Remember that "survival value" here does not mean value for a gene in a gene pool, but value for a meme in a meme pool. The question really means: What is it about the idea of a god that gives it its stability and penetrance in the cultural environment? The survival value of the god meme in the meme pool results from its great psychological appeal. It provides a superficially plausible answer to deep and troubling questions about existence. ... God exists, if only in the form of a meme with high survival value, or infective power, in the environment provided by human culture. (192–3)

Like genes within a given population's evolutionary process, memes thrive or perish insofar as their qualities are advantageous and superior to other competing memes within a particular culture. With far-reaching implications across a number of fields, Dawkins's theory of memes as replicators residing in people's brains that are able to replicate themselves as a consequence of socialization during information transmission through any media between individuals offers an explanation for cultural evolution, or rather, the gene–meme co-evolutionary process. It could also elucidate how subjective and official territories become and come-to-be, how state power, manifesting in ideas and their steady articulation, achieves predominance in given fields, and how transversal power may emerge in the fields as the result of a memetic intervention.

The field of "memetics" has proliferated since Dawkins launched his theory in 1976, and has become divided, as Robert Aunger describes it, into the "meme-as-germ" and the "meme-as-gene" camps (21). In the first camp are the memeticists, like Aaron Lynch and Richard Brodie, who take an epidemiological approach and see memes as cultural microbes that disseminate information as "mind viruses" (Dawkins' term) or "idea viruses" that get "sneezed" by people through various media into the brains of other people.[5] They exist parasitically on human hosts, making them act in ways conducive to getting copies of their information into the brains of others. Memeticists belonging to the second camp, like Daniel Dennett and Susan Blackmore, argue that memes, as gene-like replicators, account for a process of cultural evolution – how humans use language and media to circulate information that networks and substantiates societies ever more rapidly and complexly – that exceeds our understanding of what genes are capable of doing.[6] Building on the two camps, Aunger, echoing Dawkins's original position (Dawkins 323), offers a third, "electric" view on how replication of information may be both neurological and social. Aunger posits a model of what might be the physical properties of the as yet unobservable memes, which may give his theory more provable potential than the others' theories.

Generally in accord with the two major camps, I too believe that cultural traits are passed on from one individual and/or group to another through various media so that values, beliefs, and affinities are adopted both consciously and unconsciously; and that the more popular traits become, the more affective they also become, and thus the more likely they are to outlast the less popular. I am also intrigued by the shared premise of both camps that memes have agency that endeavors in the interest of their own survival, that ideas, which can replicate and distribute via interactions between humans, have the power to influence humans against their wills.[7] However, I reject Blackmore's claim that because humans are reducible to the memes and genes that program them, "There is no need to call on the creative 'power of con-sciousness,' for consciousness has no power" (236). Blackmore defines "consciousness" as "subjectivity – what's it like being me now" (238), and also maintains that, "Free will, like the self who 'has' it, is an illu-sion" (236).[8] In my view, even if subjectivity can only manifest in culture to the extent that people are made into subjects within a society, a topic discussed in Chapter 2, one must be aware (conscious) in order to, first of all, eat voluntarily; and it is from food that we have the energy to do more, or at the very least to stay conscious. Consciousness's power thus lies in its potential. It is awareness of potential in the potential consciousnesses of all people that enables transversal poetics, perhaps as a "meme machine" (as Blackmore would put it), to inspire people to become and come-to-be more aware and different from what they were. As for "the self," according to trans-versal poetics, it is always in-progress and processual, undergoing becomings and comings-to-be, and it only ever exercises freedom as the process manifests in relation to the environments through which it consciously moves.

To explain further, let me return to the wannabe surfer. Should she come to take on or exhibit attributes of the community that she did not mean to acquire or perform, especially in circumstances in which she would prefer to do otherwise, then powerful comings-to-be have occurred, and the initiate has lost some control of her performance and subjective constitution. Memeticists would say that the surfer memes are significantly colonizing her brain; the memeplex they together comprise is systematically working to marginalize other memes and break down networks that might guide her into occupa-tions that carry greater value for social structures created by other meme machines that do not privilege surfer codes in the same ways. Imagine our hypothetical wannabe surfer inadvertently blurting out the phrase "right on dude" to express enthusiasm during an interview for a corporate or academic job. Psychoanalysis might attribute this to a failure on the part of the superego or conscience, which regulates desires and thoughts such that only those appropriate to the social circumstances are outwardly expressed. In transversal terms, both becomings and comings-to-be are always at work, and in the case of the would-be surfer's verbal slippage, how the event was framed by the interviewee is the strongest indicator of which process dominated her consciousness during the event. If she did not really want the job, but only went to the interview to please her parents, then becomings-surfer won out. If she really wanted the job, perhaps to please her parents, then comings-to-be surfer took control. Of course, there are many other becomings and comings-to-be operating within this scenario, but in all cases some are privileged more than others, and the structures of some are stronger than those of others. The ratios, however, are usually commensurate with duration and profundity of engagement with alternative, captivating forces, like sociopolitical conductors and memeplexes, that have the capacity to achieve what I call "emulative authority" for certain individuals, people, or groups.

The spread of ideas and memes, and therefore the occurrence of comings-to-be, are usually more powerful and less controllable the higher the level of the activity involved, whether through performance or cognition. In other words, awareness of how to do something – knowledge of technique and instructions – is more deeply assimilated and more capable of programming behavior than simply copying the action. Blackmore refers to this phenomenon as the difference between the "copy-the-product" and the more effective "copy-the instructions" modes of transmission and replication (61–2). Learning how to act in the process of becoming a theater actor, for example, would likely instigate comings-to-be of a nature perhaps altogether different from how the lifestyle of an actor is commonly perceived, insofar as the kinds of somatic knowledge required of actors could be beneficial to people in other fields, such as surfing or physics. Watching a theatrical performance, construction work, or pornography may stimulate similar neuron firings and descriptively comparable conceptual, emotional, and/or physical responses, but it has yet to be determined whether knowledge of someone else's experience, including a character's, without having had a like experience with which to relate the knowledge can qualitatively and effectively yield the same sorts of becomings and comings-to-be that are achieved through firsthand experience.

Whatever the specific results of performing, experiencing, thinking, reading, and/or writing about any culturally dominant iconic subject – whether sports, popular music, film, or literature – and its myriad permutations, such interfacing with emulative authority engenders more profound becomings, comings-to-be, and transportation through spacetime thresholds than the quotidian activities of everyday life. When encountering or embodying any media conceptually and/or materially imbued or manifested by an icon's "affective presence," which is the combined material, symbolic, and imaginary existence of a concept/object/subject/event and its multiplicities, we become situated as participants within articulatory spaces and their overlappings and fusions, in much the same way that subsets and their elements function in mathematical set theory.[9] For instance, Shakespeare's affective presence as, among other things, marvelous poet, cultural icon, and/or ideological symbol engages us with the phenomena of what Donald Hedrick and I refer to as "Shakespace," a term that encompasses the plurality of Shakespeare-related articulatory spaces and the time, speed, and force at which they transmit and replicate, like memes, through places, cultures, and eras.[10] Other articulatory spaces explicitly engaged by my collaborators and I in the chapters of this book include Mary/Mollspace, Marlowespace, Hamletspace, R&Jspace, and theaterspace. We are especially interested in how and why the affective presences of their primary components have been inherited, varied, and selected across histories and cultures.

Fugitive explorations

In addition to clarifying further our relationships to the articulatory spaces through which we evolve, I want to demonstrate some of the ways by which the literary, cultural, and critical articulatory spaces that have significantly influenced our understandings of subjectivity in early modern England and today can provide opportunities for enhanced individual and group agency, artistic inspiration, and political and societal transformation. With these goals in mind, the "investigative-expansive mode" of analysis that guides the praxis of transversal poetics has led me to propose a corollary methodology that I call "fugitive explorations."[11]

Engaging the framework of boundless potential proposed by transversal theory, fugitive explorations call for readings of a given text – with "text" understood as anything analyzable – that defy the authorities that reduce and contain meanings, both of the readings and of the text itself. Dominating authorities can be found in all readings and reading environments, both of a text's inception and point of reception; they are the past, present, and future authoritative, interpretive communities that channel and situate a text and its interpretations across spacetime, arbitrarily producing its history and value. Hence, fugitive explorers venture wherever they are drawn (as in my journey into the field of memetics), reconstituting parameters accordingly, as they strive to uncover "fugitive elements" – human, narrative, thematic, semiotic, and so on – of the subject matter being examined and the environments in which it has been contextualized, particularly those that pressurize the authorities and, by extension, the communities necessary for the substantiation of the authorities' power. Fugitive explorers often endow agency where agency had been wanting, evacu-ated, or forbidden. Thus, in response to the fact that transversal poetics makes no overarching or definitive claims with regard to its specific political investments, but rather remains as fluid and case-specific in its determinations as the ideas, opportunities, and methodology it promulgates, I offer fugitive explorations as a derivative, transversal approach with a pronounced agenda: to understand and empower fugitive elements insofar as doing so generates positive experiences.

Politically invested, fugitive explorations might, for instance, involve an attempt to transversally link, within histories, cultures, and meta-physics, the ectoplasmic traces of the ghost-characters of Shakespeare's *Cymbeline, Julius Caesar, Macbeth,* and *Hamlet* to theatrical and literary-cultural productions of the plays. The purposes of this undertaking might be to give voice to commonly marginalized or elusive perspec-tives on or of the ghost-characters, to highlight opinions about the phenomena of ghosts across histories, cultures, and disciplines, and to converse with scientific theories that might illuminate in unexpected and productive ways the ghostly subject matters under investigation. To give some examples of fugitive inquiry from this book, Amy Cook and I follow ideas expressed in *Macbeth, Doctor Faustus,* and *The Devil is an Ass* concerning the transversal power of comedy and relate them to recent research on humor in cognitive neuroscience; Donald Hedrick and I connect anxieties expressed in *The Changeling* about the dimin-ishing importance of the concept of place to an emerging definition of space during the European and English Renaissances that privileges space at the expense of place; Janna Segal and I track recent trends in literary criticism on *Romeo and Juliet* and trace the financial subterfuge operating in the text to consider how the play simultaneously celeb-rates the production of romance and laments the title characters' investments in the business of love; and Glenn Odom and I situate *Titus Andronicus* in terms of its literary, scientific, and social history to give an expansive account of the meanings of the words "black" and "Moor" in relation to notions about ethnic, religious, and national difference within and beyond the play during the early modern period. Each of these chapters attempts to make emergent a previously over-looked aspect of the critical history on the plays and concepts discussed.

To elucidate some of the characteristics and advantages of fugitive inquiry within and beyond transversal poetics, I want to now offer a brief comparison – elaborated upon in Chapter 2 – with the philo-sophy and methodology of deconstruction as formulated by Jacques

Derrida. Like deconstruction, fugitive explorations pursue slippages, loose threads, and latent signifiers in a chosen text as a means by which to undermine and unravel the text's apparent meanings for a given interpretive community or communities. Unlike deconstruction, however, it does this deliberately as a gateway to other possible readings and, by extension, to other conceptual, emotional, and physical localities. Therefore, fugitive explorations do more than merely expose the instability of texts and the semiotic systems in which they function. It would not be enough to show how the witches in *Macbeth*, for instance, undermine through prevarication or "powers of suggestion" the patriarchal system within the play. The fugitive explorer might also relate Shakespeare's representation of witches to dissident or exploitative occasions precipitated by the circulation of seductive or misleading concepts outside of the play text – in, for example, contemporary advertising campaigns or religious institutions – as a means by which to illuminate types of becomings, comings-to-be, and subject performances that make possible, encourage, or inspire, at least conceptually, such currently hotly-debated cognitive interventions as "leading the witness" and "faith healing."[12] Along these lines, in Chapter 3, Anthony Kubiak and I tease out of the deceitful operations and multilayered performances in *Hamlet* a theory of mind that we attempt to explain in relation to pioneering research in the rapidly-growing subfield of primatology known as "Machiavellian Intelligence."[13]

Finding potentialities in instabilities, as in the cases of *Macbeth*'s witches and *Hamlet*'s artifice, fugitive explorations emphasize the text's possible meanings beyond its intended, immediate, or future audiences. A goal of transversal poetics in and through fugitive explorations is to discourage hermeneutical reductionism, such as of the kind that forces an investigation to bow down willy-nilly to overdetermined concepts like historicism, presentism, or futurism. Whereas deconstruction also opposes such reductionism insofar as it rejects ultimately all interpretations, transversal poetics asks that we consider artifacts positively and extensively, rather than define negatively, defer continuously, or dismiss alternative interpretations and applications by relying only on dialectical argumentation. Yet, like deconstruction, it also asks that we remain aware that there is no inherent, absolute, or unmediated meaning or subject position; that truth and perception are processual and contingent; and that any text or social identity (like Derrida's own writings and affective presence) can be made to deconstruct itself endlessly by systematically replacing one supplemental, always already indeterminate meaning after another, each standing in for the never-to-be-found conclusion or transcendental signified. This is done as a result of what Derrida calls *"différance,"* a coinage simultaneously meaning to differ and to defer that accounts for the fact that all meaning in a relational system of language where words do not have intrinsic meanings always both differs from and defers to other meanings, and is therefore contingent and provisional. While such deconstruction can be valuable, especially when implemented to undermine oppressive rhetoric and systems, it often leaves unanswered questions significant to people – like our students – who want to relate the literary text in question to issues pertinent to their lives.

When studying *Macbeth*, for instance, students often contemplate who is ultimately responsible for Macbeth's actions. They are usually unsatisfied when I suggest that the play is merely words, "a tale / Told by an idiot, full of sound and fury, / signifying nothing" (5.5.27–8), and that no one is responsible because there is nothing there but indeterminacy, for which anyone could be provisionally responsible. Instead, my students have defined a range of culpable agents: Duncan, the inept and careless ruler; Macbeth, the naïve, ambitious, and weak would-be king; Lady Macbeth, the power-mongering emasculator; the prophetic and interfering witches; Shakespeare, the authorial agent; the early modern English society that produced Shakespeare; and we, the immediate interpreters of the play. My point is that who or what is responsible for Macbeth's actions matters only inasmuch as we, as scholars and teachers, can productively associate the question, examination, and possible answers with issues important to people today, and – positively – to people "tomorrow, and tomorrow, and tomorrow" (*Macbeth* 5.5.19).

Hence, fugitive inquiry, working to reveal portholes, expand passages, and promote travel into disparate territories, does not pursue or resolve comfortably with the nihilism of deconstruction or with the notion of infinity that it invokes (the fugitive explorer is never a caged gerbil forever spinning on a treadmill). Nor does fugitive inquiry privilege or reduce itself to any axioms, including the influential psychoanalytic and post-Marxist axioms, discussed in Chapters 2 and 9, that define desire and subjectivity as predicated on lack. Fugitive inquiry is not a victim or victim-making approach. For transversal theory, subjectivity is processual and develops positively through becomings and comings-to-be; this often occurs through the recognition of differences, but not typically or desirably as a consequence of negation. Accordingly, compensation or totalization is not the objective of fugitive explorations, although sustaining an understanding may be a welcomed outcome. Informed by transversal theory, fugitive explorations recognize limits within circumstances and agents, even while remaining steadfastly committed to both the concept that anything is possible and the fact that there is a real where things are, happen, and can be done, however difficult to access or influence, and however subject to mediation and matters of perception. This is where fugitive inquiry most departs from deconstructionism, as well as from the poststructuralism of theorists as different and antithetical as Jean Baudrillard[14] and Judith Butler, who is discussed in Chapter 3.

Fugitive explorations can also willfully or inadvertently expose hidden elements that are disempowered because they can no longer operate covertly. For instance, once unknown enemies are identified and locatable, their deconstructionist mission, subversive potential, or state power can be weakened or diverted insofar as they can be quashed or co-opted. Consider, for example, Caliban's recognition and subsequent manipulation of Trinculo and Stephano as power-hungry, lustful humans (*The Tempest* 3.2). Either way, through the exploratory process, the investigative-expansive thinker becomes fugitive as a means by which to move transversally outside of one's own "subjective territory," the combined conceptual, emotional, and physical range from which a given subject perceives and experiences.

Although I believe that we exist simultaneously in spacetime, I use spatial metaphors to discuss subjective and other conceptual-emotional-physical territories rather than temporal or spatiotemporal ones for three reasons. First of all, and least consequentially, Western culture privileges spatial metaphors when thinking about subjectivity and social hierarchy. This can be seen in the colloquialism, "I don't want to go there," meaning a conceptual-emotional space instead of a physical place or module of time. Spatial metaphors are thus easier for people to relate to given their everyday experiences in a spatially-biased world. Another example of this bias can be seen in the fact that we often forget – become unaware of – what time it is, but rarely do we forget the space we currently inhabit. Secondly, and most importantly, I imagine territories occupying at least four dimensions – length, breadth, thickness, and time as a coordinating dimension – and having buoyancy and emergent properties such that the always malleable and developing whole can never be ossified, reduced to a sum of its parts, or disconnected from the environments it interfaces, that constitute it, and through which it moves. Thirdly, it is the complex dimensionalities of subjective territories that allow for differences and commonalities among them, and thus for the layering of their characteristics that produces and reinforces definable "official territories" that together create the infrastructures for societies, as well as for opportunities to move transversally beyond their established parameters.

Fugitive explorations can work to get people outside of their subjective territories through "transversal movements": feelings, thoughts, and actions alternative to those that work to circumscribe and maintain one's particular subjective territory, and, by extension, the greater official territory that the subjective territories of a society's members together comprise inasmuch as they share common components. Most people engage in transversal movements to some degree and in some form everyday. People most often move transversally when they empathize or imagine they are empathizing with others. As actors and therapists are often trained to do, people identify with those to whom they are empathetic, whose thoughts and feelings they may only be able to presume; one thinks and feels atypically in the attempt to empathize "as if" they are someone else, which pushes them transversally. By inhabiting, if only imaginatively, the subjective territory of another, one's own subjective territory expands and reconfigures, and the higher-order, stabilizing official territory may be jeopardized if the emotional-conceptual motion is incongruous with or fugitive to its biologically and ideologically informed network of organizing principles.

The transversal inclination is always fugitive to the subjectified, but not all transversal movement is fugitive on every level. Someone occupying conceptual, emotional, and/or physical spacetimes alternative to those prescribed by an official culture has moved out of her subjective territory, is expanding her experiential range, possibly disidentifying with her established social role, and is acting fugitively in a self-referential way. If someone's transversal movement works in the interest of dominant sociopolitical conductors – and thereby promotes the overarching "state machinery" that the conductors together comprise[15] – to institute a subjective territory that reinforces official culture, the dissident potential of the fugitive action can diminish. This might happen if a member of a criminal group violates the group's codes, thereby becoming fugitive to the group, and inadvertently supports the mainstream culture in relation to which the group defines itself. Whatever the outcome, the person's transversal movement may nonetheless serve as a model that inspires others to wander.

Theaterspace, paused consciousness

Theater happens when a performance is presented to an intended audience that is aware of an interpretive frame specific to that performance. As an event, in effect, of performance, theater is often an exemplary model of the kind of apparatus that induces transversal wanderings through processes of becomings-other and comings-to-be-other, such as of other social identities, species categories, or spiritual beings. Theater spurs these adventures through such phenomena as empathy, projection, hypothesis, and transference. As I have demonstrated elsewhere, this was especially the case for early modern England's public theater.[16] The metatheatrics and impersonations that characterized the public theater, unlike allegorical presentations in pageants and morality plays, challenged established beliefs about the singularity and cogency of reality and, by extension, the sexual, gender, moral, and class differentiations that depended on those beliefs. The public theater posited all social categories, such as "man" and "woman," as constructs that must be performed in order *to be*. It suggested that the body matters only to the extent that it is a point of departure for identity becomings, and that through performance one could potentially become anything, even the unthinkable.[17]

The public theater's transversal influence in early modern England – and the theaterspace engendered there – is especially evident in the vehement antitheatrical discourse and actions taken against the theater. To a lesser extent, it can also be seen, as Ayanna Thompson and I argue in Chapter 7, in Marlowe's creation of a new theatrical form we term "post-theater." Yet, the transversal influence of the theater was most manifest in the workings of criminals and social deviants, such as individuals who disguised themselves as gypsies in order to extort money, sell herbal remedies, and read palms; con-men who pretended to be different people in order to perpetrate crimes; and people who practiced transvestism, whether male-to-female on the stage or female-to-male on London's streets, the latter having become a popular fashion contemporaneous with the public theater's popularity. The transversal power of the public theater transgressed the Church's ideology and corresponding official culture, undermining the properties of the society's interiority that functioned to organize and monitor via sociopolitical conductors the subjective territories of its members. Thus, as result of the curious situations, unexpected events, and surprises staged in and effected by the public theater that made spectators more susceptible to transversal power than they might have otherwise been, audience members had experiences alternative enough to their subjective and official territories that they assimilated unfamiliar information and codes – maybe otherwise unwanted memes – and became and came-to-be things different from what they were before they entered the theater. Most importantly, they acquired the radical information that change of all kinds, physical as well as social, is possible.

Surprise, meaning an unanticipated experience outside of the parameters within which one normally experiences life, can effect transversal movements with tremendous intensity and impact. Moreover, if one considers that the more improbable an event for any synthetic or ecological system, the greater that event's potential consequences for the system, surprise can vary in degree; it can even be exponential (think of the 2004 tsunami in Asia, and what it was like as new information about casualties and damage was revealed). This is especially the case when surprise is combined with loss of control. If surprise, as typically defined, means to be taken unawares, and since consciousness, as typically defined, requires awareness, surprise suggests at least a temporary loss, disruption, fracturing, or what I call "paused consciousness." With this in mind, I would like to discuss briefly why humans wish to experience theater.

Although there are many reasons why people attend and participate in theater, I want to concentrate on a few that I believe are primary and particularly significant. People often measure the quality of their theatrical experience by the extent and depth of movement experienced. In other words, should one attend a production framed as comedy, one expects to smile and laugh, and, hopefully, in the best of circumstances, shed tears from excessive laughter. Successful jokes require breaks from established discourses and frames through deviations that take the listener elsewhere, to the unexpected. Limits for jokes are usually previously determined and longstanding for a given social group, and so a joke that goes "too far" is one that violates the frame of the joke itself. Such violations take the listener to a more precarious, alternative space from which it is more difficult to return to the conceptual-emotional place of the social situation from which the joke initially departed. The violation suggests that what has been expressed is not a joke after all, but instead, say, an insult, truth, or a divisive idea possibly motivated by ill or revolutionary intentions. Nevertheless, because the joke violated social code, it is a threat to the organizing structures of subjective and official territories.

On the other hand, when one attends a theatrical production framed as tragedy, one expects to feel sad, perhaps even to cry. Joy and sadness, then, are experienced in theater even though the actions presented are usually not those of the audience: representation stimulates real responses often considered appropriate to corollary, offstage, everyday events. One reason for this, as, among others, Maksim Stamenov and Vittorio Gallese have demonstrated, is that the same neuron firing among brain cells called "mirror neurons" occurs when we imagine or watch an action we are not actually performing (11).[18] The intensity of firing, and therefore experience, may differ depending on already established pathways for neuron firing specific to each individual as a result of previous knowledge and experience with the action;[19] generally, as in Blackmore's "copy-the instructions" model, the more familiar the individual is with the action, the greater the rate and potency of firing. Evidence suggests that we can experience confusion, anxiety, or pleasure when neurons fire in drastically atypical or random ways, that is, when common thought patterns yaw or rupture. Thus, despite the specific genre frame – comedy, tragedy, tragicomedy, history, romance, and so on – theater is most enjoyed or disliked when it exceeds our expectations, when it is more of something – "worse," "better," or "weirder" – than anticipated, which is when comings-to-be are most likely to transpire. Of course, a production's frame contributes directly to the construction of our expectations and the conditions for their failings or successes.

At the subjective crux of such occurrences, however rare, ominous or delightful, are the secrets that we – and I am using "we" proverbially, not presumptuously – consciously and/or unconsciously carry within us. These secrets become, if only provisionally, exposed through our reactions: they are our ready gateways to alternative thinking and becomings, our fugitive inconsistencies – in sensibility, identity, and cognition – always susceptible to the power of the transversal to unearth, liberate, and transform. In the reflections of what we experience, we become vulnerable to our own scrutiny, and perhaps to the often more threatening scrutiny of others. This happens through imagination, empathy, projection, and transference – by getting outside of one's subjective territory – rather than through self-affirming identification, defensive reactions, or alienation, although defamiliarization, a foreign perspective on something familiar, can be effective, as epic theater has shown. But it also happens, in conjunction, through excess.

When we laugh so hard that it hurts, the comedy is considered successful. When we cry so hard or feel so sad that it hurts, the tragedy is deemed successful. Clearly, we want to hurt: to lose control and feel wildly, profoundly. We respond involuntarily, free from responsibility. Since we generally do not inflict pain directly on ourselves, as this would be seen as irrational, we go somewhere to have it done to us. Among other places – the movies, sports events, the homes of relatives – we go to theater. When we leave the theater with our bodies aching and our brains reeling, unable to contain our thoughts and feelings, we feel that we have got our money's worth. Time, a precious resource, was well spent if the duration of experience persists beyond the immediate, actual spacetime of the spectacle. If our philosophical or ethical meditations on the performance just seen permeate our lives for hours or days afterwards, despite our efforts to focus on other things, we have been possessed and are no longer in control of their influence.

This lasting impact is especially felt when theater scares us, tells us of the possibility of, say, fatal attractions or serial killers, or that everything in our lives – our family, friends, ourselves – may be, like theater, simply artifice, ready to betray us unexpectedly, and perhaps devastatingly. If theater causes us to lose control through laughter, squirming, terror, and tears, through bodily contortions and incontinence, then it has done its job; it has moved us transversally. In fact, if we forget where we have been – in space as well as time – during a theatrical performance, and sort of snap out of a dream when it is over, then we have experienced paused consciousness. The event is more memorable, more effectual, even if or because the details are fuzzy. This is because we have been possessed, prodded, made to consider alternatives, even the otherwise inconceivable, as we try, often with urgency and desperation, to fill in the gaps, to impose order and sense onto mystery (state power at work). Nothing in theater, as in everyday life, pushes us more than transversal power manifested as, through, and by surprise, deceit, irony, and the "subjunctivities" that question, challenge, and transform our subjective territory.

Empathy, surprise, lying, and imagination are the most common means by which people venture into what I call "subjunctive space," the hypothetical worlds of both "as ifs" and "what ifs" that interface subjective territories and what I have termed "transversal territory," a multidimensional spacetime encompassing, among other known and unknown qualities, the nonsubjectified regions of individuals' conceptual-emotional range. Subjunctive space, because of its openness and uncertainty, is like transversal territory, whose indeterminate mappings can occupy, transgress, and expand – to borrow Raymond Williams' terminology – the "residual," "dominant," and "emergent" aspects of a culture.[20] In this respect, subjunctive space is a subset of transversal territory or, more precisely, it is an emergent space operative in between and among subjective and transversal territories in which the subject necessarily retains agency and self-consciously hypothesizes scenarios and experiences, self-activating his or her own transversal movements. Resistance to subjunctivity, whether achieved consciously or not, can be likened to complacency, the acceptance of a prescribed subjectivity.

Consider the power of words, as in Shakespeare's *Titus Andronicus*, when Aaron tells Titus that if he cuts off his own hand, the lives of his sons will be spared. Without daring to move subjunctively, Titus believes the lie because of the Roman code to which he adheres. His subscription to the code supports the official territory within which Titus obediently operates. As Glenn Odom and I discuss in Chapter 8, it is only when Titus is surprised by the truth of his sons' murders that he derails transversally. Similarly, Iago says things that inculcate Othello into Iago's "cuckold ideology" and lead Othello to murder Desdemona without due process, as Venetian law demands.[21] To move away from Shakespearean examples and into everyday life, consider the things people do merely because someone has told them, "I love you." In all cases, the greater the investment in the words by the audience, given the significance of the speaker(s) to the spectator(s), the greater the liability. Saying, "I promise to take out the trash," is not equivalent to saying, "I promise to feed the baby," just as performing or not performing the action can yield very different results. When codes and frames are broken, we respond accordingly. We value some actions more than others, and so the consequences are greater. Subjunctive space allows us to consider events and their consequences differently, even fugitively, in effect creating opportunities for thought and expression where there were none.

Fugitive subjunctivity

Whether through theater or critical inquiry, fugitive explorations can be precarious undertakings because they challenge, defy, and promote the defiance of authorities, which can lead to real consequences, including social metamorphosis, combat, punishment, and liberation. Transversal movements, moreover, are often corollary to fugitivity, and can take one deep into transversal territory, which could likewise cause the wayward traveler to experience a cognitive disjunction that is often pathological, taking the form of a dream-like state of altered

consciousness in which themes are lost and reappear, possibly endlessly (the waking nightmare or living dream of the deconstructionist).
Such persistent occupation of transversal territory can be counterpro-
ductive if the goal of the fugitive explorer is to emancipate readings of
the text, herself, and/or others in order to achieve agency. Certainly
Macbeth's irresistible transversality, influenced by his quest for what
he cannot know, pushes his subject performance into what zooz terms
a "progressive quagmire":

> Progressive quagmires are research states, indeed states of being
> (if you will), that are manifest when the analytical tools which were
> believed to fuel progress prove unable to resist the analysis' momen-
> tum and thus are incapable of generating new directionality and
> expansion because the analysis is pushed along a rigid course. And
> yet the researcher, having experienced past "successes" with them,
> and being urged on by social, cultural, political, and paradigmatic
> conventions, is reluctant to part with these investigative tools and,
> by extension, the (de)limiting assumptions underlying them. (13)

To avoid progressive quagmires by becoming fugitive analytically,
and therefore be able to ferret out fugitive components of a system,
narrative, psyche, and so on, one frequently has to journey into the
hypothetical dimensions of subjunctive space. Because the fugitive is
mysterious, perhaps already on the run, elusive, and/or burrowing in
the nooks and crannies of discourse, the transversalist engages subjunctively in atypical possibilities for meaning and articulation: the
"what ifs" and "as ifs" that the text (or experience) may or may not
inspire. Taking you with me into subjunctive space, I would like to
turn to a brief example, a fugitive dabble of sorts, into the early
modern English discourse on equivocation.

Early modern England's dominant religious ideology maintained
that God orders and Satan confuses, a determination that reverberates,
for instance, in Macduff's response to the revelation of King Duncan's
murder: _"Confusion now hath made his masterpiece!" (*Macbeth*
2.3.65). This God–Satan dynamic resounds throughout *Macbeth* as
equivocation (in language and action) and is associated with witchcraft
and the Jesuit conspirators who attempted to blow up King James and
Parliament in the Gunpowder Plot of 1605. In particular, Shakespeare's
use of the word "equivocation" by the Porter (2.3.8–9; 2.3.30–2) references *A Treatise of Equivocation*, written by early modern England's
clandestine Jesuit leader, Father Henry Garnet, which describes the

language and gestures one can use to provide deceptive answers under oath without retribution from God.[22] The general employment of equivocation in *Macbeth*, especially by the witches, who "draw" Macbeth into "confusion" (3.5.29), mislead him with riddles – "none of woman born" (4.1.79), "Birnam wood" (4.1.93), "palter with us in *double* sense" (5.8.20) – demonstrates through discourse and performance the presence of Satan within the play's world and beyond as it aligns the play's witches with both offstage witches and the Jesuits, who are referred to as "devil-conjuring priests" in Samuel Harsnett's *A Declaration of Egregious Popish Impostures* (1603) (149).[23]

Witches and Jesuits, like all of Satan's underlings, were seen as infiltrators working to promote chaos in God's otherwise orderly universe. As discussed in Chapter 4, since equivocation encourages doubt, and doubt signifies evidence of Satan's work and further opportunity for Satan to undermine order and goodness, equivocation was seen as a threat to cosmic structures, moral authority, and societal coherence, which is to say, state power. As documented by Edward Casey, according to medieval and early modern England's prevailing ideologies, God both created the universe and occupies particular places within it, and to challenge this idea, even under the auspices of hypothesis, was considered heretical and treasonous. In fact, early moderners believed that God occupies some individuals and places more than others, such as pious people and churches, with God's presence being measured by the "goodness" found in them. But if we consider alternative theories of metaphysics and subjectivity emergent toward the end of the sixteenth century and developing throughout the seventeenth, as articulated, for example, in the logic behind the Jesuits' use of equivocation, the extent to which the notions of dissemination and particularity were commonly held becomes questionable.

While researching with my collaborators, especially Donald Hedrick and Amy Cook, representations of deceit in early modern English discourse on performance, my fugitive explorations, moving investigative-expansively beyond *Macbeth* and the other texts it commonly invokes, led me to a hitherto unaddressed connection between the discourse on equivocation and an unprecedented change in perceptions of place as a possible location for either Godly or Satanic interventions. Of the historical sea changes of the sixteenth and seventeenth centuries, including the Copernican system, the great vowel shift, the public theater, nascent capitalism, and colonial expansion, this important conceptual revolution, discussed in Chapter 5, has often been ignored, although it connects all of these revolutions. While the transformation of the concept of place in relation to the concept of space was complex, variously expressed, and occurred over several centuries, it has been generally charted in the history of ideas by Casey as a move toward a de-emphasis of the idea of place as it had been construed in Aristotelian/Ptolemaic and Christian philosophical traditions. By the early modern period, place began to lose its status, indeed its affective presence, as it was subsumed by the notion that God is infinite space, as suggested when Hamlet speaks of divine kings occupying "infinite space" (2.2.255). It was no longer thought that God occupies particular places within the infinite space that he created, but rather that his presence was infinite. He was now thought to be in all places at all times, thereby making a concept of place irrelevant inasmuch as humans are made in God's image, for if God could not occupy a discrete subject position from which to observe the universe he created, neither could humans.

Inspired by Casey's work on place, I searched for references to time and space in early modern England's commercial literature and noted that England's early moderners, predating Sir Isaac Newton's published understandings of space and time as absolute states, conceived of people as existing in the infinite space that is God, which also accounted for infinite time. For instance, Shakespeare often makes time and space synonymous, as in *1 Henry VI*, where things happen "after three days' space" (3.2.294), or in *Love's Labor's Lost*, where the action spans "three years space" (1.1.52). Because people equated space with God and God with time, seeing them as the same substance, what makes equivocation radical in relation to this revolutionary conceptual change is not just that the ambiguity it produces allows for Satanic interventions. The acknowledgment of the everywhereness of God in the logic behind the employment of equivocation – of God occupying placelessness because he is simultaneously in all places – that anticipated and contributed to the succession of space over place that gave way to the scientific revolution also implies a radical understanding of subjectivity. A causal, reciprocal relationship between this new idea of God's pervasiveness/placelessness and the idea of an open-ended phenomenology ironically problematizing the Cartesian subject – who knows he exists because he thinks, but only knows this because God is the source of truth who bestows thought – is supported by the theories of Galileo Galilei and later crystallized in Newton's third law of motion, in which the actions of two bodies upon each other are always equal and directly opposite. Unfortunately for some people, such as the seventeenth-century Dutch philosopher Baruch Spinoza, who was less willing to veil his subscription to this fugitive concept of relational effect than Newton was, the fact that this law leaves no room for Satan, or, inversely, for God, resulted in their displacement. Indeed, the radical ramification of the all-places/all-times model of God is that it leaves no room for the concept of autonomous and/or individuated subjectivities. Since it posits subjectivity as always pluralized, infused, and collective, subjects appear unique and differentiated only insofar as they convince each other that they are individuals through framing and performance, which is to say, through theater.

Emergent activity

For me, consciousness emerges from properties of matter, but is not reducible to any specific, observable matter of which we are; that is, consciousness is irreducible to the extent that its physical properties have yet to be discovered. Consciousness performed, and therefore measurable, is interlaced and conditionally and effectually related to mental phenomena, like intentionality, subjectivity, and desire. There are conscious and unconscious states, as well as paused consciousness, when self-awareness is suspended while the individual is "awake." Like John Searle and others, I believe that consciousness refers to the experience of being aware and sentient.[24] However, I also subscribe to the more controversial idea of multileveled, higher-orders of consciousness (although higher-order access may not unravel the mystery of consciousness).[25] In other words, we can be aware of our consciousness from another conscious standpoint, much as theater makes us aware of how we script, perform, and construct our lives. I also think that we are always engaged in becomings and comings-to-be among different mental states, as when we drive a car on "autopilot," that is, when intending to drive elsewhere, we instead automatically drive to a destination we habitually drive to at a certain spacetime. Such conditioning can be seen too in the habitual use of methodologies in literary-cultural criticism. Consider also when we move between REM sleep and lucid dreams; this is mostly an unconscious experience, often oscillating between becoming and coming-to-be conscious.

Moreover, apperception or perception of oneself or selves in the process of developing cognitively is characteristic of consciousness, thus making possible what I call "reflexive-consciousness," which connects directly to one's subjective territory. Subjectivity is phenomena not only interactive with other mental states of an individual, but also with the world outside of an individual's body: mind, body, and environment are connected even though they often appear to operate discretely. One of the ways by which people exercise reflexive-consciousness or higher-order consciousness is through subjunctive contemplation of the "as ifs" and "what ifs" that work to situate an individual's subjectivity and experience in relation to the past, present, and possible future. Subjunctivity is crucial to more sophisticated man-ifestations of consciousness, as is irony and humor, which can be seen in conceptual plurality and nonlinear conversation. I believe that consciousness – with its elementary properties existing in matter – is located in the brain in relationship with the external world when actively engaged with it, whether in actual or imaginary terms, which are determinations often impossible to differentiate. Since no reality comes to us without mediation, consciousness cannot either; it may be in the interest of consciousness to remain mysterious. Subjective territ-ory, then, is a byproduct of consciousness's interaction with the collective or clustered consciousnesses of other people that work together to organize around a particular intentional stance or position-ality, and this is how official territories and societies form through state power. As an organizing principle, consciousness gathers around indi-viduals, as well as within and around groups. Consciousness organizes complexity, is created as a result of complexity, and can be discom-bobulated in effect of new complexities, which can emerge through the subjunctivity and transversal movements inspired by such con-sciousness-reflecting phenomena as theater.

This collaboratively authored book, a phenomenon structurally and methodologically similar to both theater and consciousness, is an example of what William West and I refer to as an "emergent activity": its various findings, investigations, and arguments are irreducible to the minds, bodies, and environments that became and came-to-be each other, often undergoing "rematerializations," in the process of creating it.[26] My collaborators and I have organized the book through, around, and in response to the interfacing articulatory spaces of trans-versal poetics, fugitive explorations, and early modern English studies, trying to manifest them together with emulative authority, reflexive-consciousness, and investigative-expansiveness in the interest of fostering learning, conversations, and development of thought and experience. Thus, we seek to replicate the ideas that take precedence within the book, as memeticists would expect, beyond its individual chapters and the more coordinated expression that the chapters together comprise. Yet we do this to actualize rather than fix the ideas, to unleash organically rather than ossify them, to give others the opportunity to examine, correct, adapt, or discard them. As in *Performing Transversally: Reimagining Shakespeare and the Critical Future*, my collaborators and I have included some repetition of explanations of terms and theories in each of the chapters. We have done this so that the chapters may be read independently from the rest of the book, such as in college courses or by scholars researching the subject of just one chapter. Nevertheless, we hope that readers, intrigued by what they encounter in one chapter, venture investigative-expansively into others. This was our experience as we wrote this book. We set out to explore certain topics and ended up broadening our anticipated fields of inquiry, incorporating unexpected variables, even stumbling across fugitive elements that shifted the sands from which we embarked. With our fugitive explorations, we have attempted to identify and analyze transversal enterprises within the drama of Shakespeare and his contemporaries so that others may continue the journey.

Notes

1. See Bryan Reynolds, "The Devil's House, 'or worse': Transversal Power and Antitheatrical Discourse in Early Modern England" (*Theatre Journal* 49.2 [1997], 143–67); *Becoming Criminal: Transversal Performance and Cultural Dissidence in Early Modern England* (Baltimore: Johns Hopkins University Press, 2002), 1–22; and *Performing Transversally: Reimagining Shakespeare and the Critical Future* (New York: Palgrave Macmillan, 2003), 1–28.
2. See previous note.
3. See note 1.
4. For an analysis of the major theories on memes, see Robert Aunger, *The Electric Meme: a New Theory of How We Think* (New York: The Free Press, 2002).
5. See Richard Brodie, *Virus of the Mind: the New Science of the Meme* (Seattle: Integral Press, 1996); Aaron Lynch, *Thought Contagion: How Belief Spreads through Society* (New York: Basic Books, 1996); and Seth Godin, *Unleashing the Ideavirus* (New York: Do You Zoom, 2000). For an analysis of their positions, see Aunger, 7–32.
6. See Daniel Dennett, *Consciousness Explained* (Boston: Little, Brown and Company, 1991), and Susan Blackmore, *The Meme Machine* (Oxford: Oxford University Press, 1999). For an analysis of their views, see Aunger, 7–23.
7. See my similar theory of "objective agency" in *Becoming Criminal*, 24–7. However, I do not argue that objects vie, like genes or memes, for their own survival.
8. See Daniel Dennett, *Darwin's Dangerous Idea* (London: Penguin, 1995), 365, and *Consciousness Explained*, 210.
9. See Reynolds, *Becoming Criminal*, 1–22; and *Performing Transversally*, 1–28.

10. See Donald Hedrick and Bryan Reynolds, "Shakespace and Transversal Power," *Shakespeare Without Class: Misappropriations of Cultural Capital*, eds. Hedrick and Reynolds (New York: Palgrave Macmillan, 2000), 3–47; and Reynolds, *Performing Transversally*, 1–28.
11. See Reynolds, "The Devil's House"; *Becoming Criminal*, 1–22; and *Performing Transversally*, 1–28.
12. See Reynolds, "Untimely Ripped: Mediating Witchcraft in Polanski and Shakespeare," in *Performing Transversally*, 111–36.
13. See Richard W. Byrne and Andrew Whiten, eds., *Machiavellian Intelligence: Social Expertise in the Evolution of Intellect in Monkeys, Apes, and Humans* (Oxford: Oxford University Press, 1988), and their *Machiavellian Intelligence II: Extensions and Evaluations* (Cambridge: Cambridge University Press, 1997).
14. See Jean Baudrillard, *Simulations* (New York: Semiotext(e), 1981), in which he argues that the creation of copies without originals, presented as copies that stand in for the never-again accessible originals, is a defining characteristic of postmodernism. He maintains that at this historical moment, unlike during other periods in history, the real is only that which can be or is always already reproduced, and therefore "hyperreal": there is no difference between reality and its representation; there is only simulacrum. He also discusses the early modern period, in which he claims there was only awareness˙of the counterfeit of the real in relation to knowledge of the real; images were seen as just illusions. As I discuss later in this chapter and in Chapter 4, and as many scholars have noted, this was not what many people in early modern England thought – especially teachers at Oxford and Cambridge – about the workings of Satan and, by extension, the public theater. To bring Baudrillard's thesis back to the present, consider telling somebody who just lost a loved one or an eye that their experience is only an unreal copy with no original.
15. I coined the term state machinery for a society's governmental assembly of conductors as a corrective to the political philosophy of Louis Althusser that has informed much recent Marxist scholarship, particularly that of cultural materialists and new historicists in the field of early modern English studies. With state machinery, a term that simultaneously connotes singularity and plurality, I have adapted Althusser's conception of what he calls the "Repressive State Apparatus," which includes the governmental mechanisms that strive to control our bodies, and have fused it with his subsidiary "Ideological State Apparatuses," the inculcating mechanisms that strive to control our thoughts and emotions. My purpose is to emphasize that a society's drive for governmental coherence is always motivated by assorted conductors of state-oriented organizational power that are at different times and to varying degrees always both repressive and ideological. This is a sociopower dynamic in which various conductors work, sometimes individually and sometimes in conjunction with other conductors, to substantiate their own positions of power within the sociopolitical field. Hence, my use of the term state machinery should make explicit the multifarious and discursive nature of state power, and thus prevent the misperception of the sociopower dynamic as the result of a conspiracy led by a monolithic state. This is not to say, however, that conspiracies do not occur

and take the form of state factions. On the contrary, this must be the case for the more complex machinery to run. See Reynolds, "The Devil's House," 143–67; *Becoming Criminal*, 1–22; and *Performing Transversally*, 1–28.

16. See Reynolds, "Antitheatrical Discourse, Transversal Theater, Criminal Intervention," in *Becoming Criminal*, 125–55.
17. Reynolds, *Becoming Criminal*, 150.
18. See Maksim Stamenov and Vittorio Gallese, eds., *Mirror Neurons in the Evolution of Brain and Language* (London: John Benjamins Publishing, 2002); Gallese et al., "Hearing Sounds, Understanding Actions: Action Representation in Mirror Neurons," *Science* 297 (2002): 846–8; and B. Calvo-Merino, D. E. Glaser, J. Grezes, R. E. Passingham, and P. Haggard, "Action observation and acquired motor skills: an FMRI study with expert dancers" (*Cerebral Cortex* 15 [2005], 1243–9).
19. Alternatively, or concurrently, this may have to do with the institution of what Antonio Damasio calls "somatic markers," which are emotion-generated mental markers that influence our decisions. See Damasio, *Descartes' Error* (New York: Putnam, 1994), and *The Feeling of What Happens* (New York: Harcourt, 2000).
20. See Raymond Williams, *Marxism and Literature* (Oxford: Oxford University Press, 1977).
21. See Joseph Fitzpatrick and Bryan Reynolds (with additional dialogue by Bryan Reynolds and Janna Segal), "Venetian Ideology or Transversal Power?: Iago's Motives and the Means by which Othello Falls," in Reynolds, *Performing Transversally*, 55–84.
22. In "The Intelligent Man's Guide to Lying Under Oath" (*Inside Liberty* 8.3 [1999]: 1–2), David Kopel summarizes:

    A *Treatise of Equivocation* ... offered four techniques of equivocation: ambiguity (answering "a priest lyeth not in my house" could mean that the priest hidden in the home did not tell lies); incomplete answers ("I went to his house for dinner," omitting that "I also went to attend a secret mass"); hidden gestures and pronoun references ("I did not see anyone go that way," while pointing the other way with one's finger hidden in a pocket); and the most sensational technique: responding to questions both verbally and mentally; a Catholic could "securely in conscience" provide answers with a "secret meaning reserved in his mind." If an English government attorney interrogated someone suspected of being a priest named Peter, the attorney might ask, "Is your name Peter?" A *Treatise of Equivocation* instructed that the priest could speak the word "No" in response. The priest could then continue, speaking in his own mind but not out loud, "so as I am bound to utter it to you, since you have no lawful jurisdiction over me. (1)

23. For a discussion of the relationship between witches and Catholics according to early modern England's official culture, see Jonathan Gil Harris, *Foreign Bodies and the Body Politics: Discourses of Social Pathology in Early Modern England* (Cambridge: Cambridge University Press, 1998), 118–27. As Harris shows, "The conflation worked both ways. If Catholics were identified with witches, witches were also commonly identified with Catholics" (123).

24. See John Searle, *The Mystery of Consciousness* (New York: Review of Books Collections Series, 1997), *Minds, Brains and Science* (Cambridge: Harvard University Press, 1984), and *Mind: a Brief Introduction* (Oxford: Oxford University Press, 2004).
25. See Peter Carruthers, *Language, Thought and Consciousness* (Cambridge: Cambridge University Press, 1996), and *Phenomenal Consciousness: a Naturalistic Theory* (Cambridge: Cambridge University Press, 2000); Dennett, *Consciousness Explained*; and David Rosenthal, "How Many Kinds of Consciousness," in *Consciousness and Cognition* 11.4 (2002): 167–85.
26. On emergent activities and rematerialization, see Bryan Reynolds and William West, "Shakespearean Emergences: Back from Materialisms to Transversalisms and Beyond," *Rematerializing Shakespeare: Authority and Representation on the Early Modern English Stage*, eds. Reynolds and West (Basingstoke: Palgrave Macmillan, 2005).

Works cited

Aunger, Robert. *The Electric Meme: a New Theory of How We Think*. New York: The Free Press, 2002.
Blackmore, Susan. *The Meme Machine*. Oxford: Oxford University Press, 1999.
Brodie, Richard. *Virus of the Mind: the New Science of the Meme*. Seattle: Integral Press, 1996.
Casey, Edward. *The Fate of Place*. Berkeley: University of California Press, 1999.
Damasio, Antonio. *Descartes' Error*. New York: Putnam, 1994.
—— *The Feeling of What Happens*. New York: Harcourt, 2000.
Dawkins, Richard. *The Selfish Gene*. Oxford: Oxford University Press, 1976.
Dennett, Daniel. *Darwin's Dangerous Idea*. London: Penguin, 1995.
Harsnett, Samuel. *A Declaration of Egregious Popish Impostures*. 1603; STC 12880.
Lynch, Aaron. *Thought Contagion: How Belief Spreads through Society*. New York: Basic Books, 1996.
Shakespeare, William. *Hamlet*. Ed. Harold Jenkins. London: Methuen, 1982.
—— *Macbeth*. Ed. Kenneth Muir. London: Methuen, 1951.
—— *The First Part of King Henry VI*. Ed. Andrew S. Cairncross. London: Methuen, 1962.
—— *Love's Labor's Lost*. Ed. Richard David. London: Methuen, 1968.
—— *The Tempest*. Ed. Frank Kermode. London: Methuen 1954.
zooz, "Transversal Poetics: I. E. Mode," *GESTOS: Teoría y Práctica del Teatro Hispánico* 18.35 (2003).

2

The Reckoning of Moll Cutpurse: Transversal Reimaginings of *The Roaring Girl*

Bryan Reynolds & Janna Segal

Critical contexts with a *différance*

Thomas Middleton and Thomas Dekker's *The Roaring Girl* has been among the most discussed of the non-Shakespearean early modern English plays since the mid 1980s. The reason for this may seem fairly obvious to those familiar with trends in both literary-cultural criticism and popular culture since 1980. As a play about a powerful woman cross-dressed as a man, it was especially ripe for second-wave feminist literary-cultural criticism and as subject matter for the emergent field of gender studies. After the publicized advent of the AIDS epidemic in the early '80s, attention to gay culture increased dramatically, quickly becoming common discourse in media ranging from the nightly news to cinema to academic research. However, the focus of this attention in scholarship was often not AIDS, but identity formation and the sociocultural politics and subject positioning that informed it. Phenomenological and formalist approaches waned in popularity as poststructuralism, particularly the move to simultaneously historicize, de-essentialize, and relativize cultural products (from literary texts to human subjectivities), became increasingly predominant. All meaning, in such forms as categorizations, determinations, and interpretations, was seen as ideologically constrained and specific to its social, cultural, and historical situation; the catchwords "discursive," "representation," "construction," "appropriation," "blurring," "crossing," "passing," "mutability," and "indeterminacy" became common parlance in literary-cultural criticism.

The idea that social identity (sexual, gender, class, ethnic, racial) was not inherent and was imposed, composed, and/or had to be performed convincingly enough that one "passed" (that is, would be identified positively) became common denominators to the leading critical methodologies (psychological, materialist, deconstructionist), even when applied to various areas of interest, from literary texts to theatrical characters to living cultural icons. The longstanding nature/nurture dichotomy changed nomenclature, transforming into essentialist/constructivist, and eventually giving way to the rhetoric of the constructivist position. As a result, for many critical thinkers, social identity was (and is) articulated processually, the temporal outcome of which remains socially, culturally, and historically contingent and negotiable. Insofar as sexuality and gender, and their links with race, ethnicity, class, and culture, are guiding principles behind most human relationships, the chief preoccupations of this movement in critical inquiry were sexual-gender potentiality and its affect on conventional class demarcations.

The main concern of this new focus for those either supporting or opposing majoritarian cultural traditions was the fact that anything that challenges dominant ideological perspectives on sexual, gender, ethnic, race, and class difference threatens a society's network of familial, educational, religious, and juridical structures. In the language of transversal poetics,[1] these structures are referred to as "sociopolitical conductors," a term which Reynolds coined as a corrective to Louis Althusser's ahistorical concepts of "Ideological State apparatuses" and "Repressive State apparatuses."[2] Reynolds' term accounts for the historically specific sociopolitics and conditional states of both dominant and counter-hegemonic machinations for prescribing ideologies, whereas Althusser reduces these conductors to a binary system of a monolithic state, and simultaneously does not acknowledge the similarly functioning, often multiple and contradictory ideological machinations of non-state promoting entities.[3] While sociopolitical conductors reinforcing a given society are mutually dependent on dominant ideological perspectives, other sociopolitical conductors operating within or beyond the society may oppose the dominant perspectives, either wholly or in part. The amalgamation of a society's state-serving sociopolitical conductors must, in the interest of fostering a "rational" social system, support and generate "state power," which are forces of coherence, whether acted consciously and/or not by or upon "individuals" (humans individuated from other humans) or "groups" (a social clustering of otherwise individuated humans). In transversal terms, the amalgamation is referred to as "state machinery,"[4] a concept that accounts for the singular and/or plural, human and/or technological influences that work tirelessly but ultimately futilely to manufacture moral, ideational, and governmental coherence and symbiosis. According to transversal theory, hegemonic work is, in the end, futile if totalization is its goal because totalization is a social impossibility as long as physical movement and change are constant realties. Consequently, an absolute state for both individuals and/or the society they comprise is never a real prospect, but the quest for stable states and the fear of achieving them will nevertheless always stimulate solidarities and antagonisms among sociopolitical conductors with various views on what the ideal society and state should be.

Such a reality serves the non-economist, neo-Gramscian post-Marxists Ernesto Laclau and Chantal Mouffe,[5] who emerged influentially on the academic scene in the '80s with the publication of their *Hegemony and Socialist Strategy* (1985), and continue to maintain their same views into the twenty-first century.[6] The world's constant flux is compatible with their poststructuralism insofar as they imagine an overarching hegemony, "a collective will," constituted through the "politico-ideological articulation of dispersed and fragmented historical forces" that can only and must be resisted through antagonism (67). The ceaseless production of antagonism is therefore at the "centre" of their mission to keep what they identify abstractly as the political "Left" and "Right" from meeting harmoniously, presumably on any issues, in the "Centre" of what they believe is a linearly-manifest continuum of political thought and activism (xiv–xv). Strategically, Laclau and Mouffe's "new left-wing hegemonic project" stresses the need for antagonism as instigated through the "recognition" of value and "redistribution" of power as best-suited to their greater purpose, the pursuit of a "radical and plural democracy" which cultivates endlessly the extension of long-established democratic struggles for representation, equality, and liberty to and within marginalized class, race, ethnic, gender, and sexual identities (xviii). Their insistence on an indefatigable "hegemonic struggle" (xix), combined with their resistance to classical Marxism's totalizing and deterministic sociohistorical schema, clashed with the precursory and more influential historical materialism of Fredric Jameson.

Jameson argues in his Althusserian-Freudian-Marxist merger, *The Political Unconscious* (1981), that the mediation of differences in identity and subjectivity, among other sociohistorical phenomena, are, while often convincing, misleading, or confusing, finally unable to detract from the fact that "social life is in its fundamental reality one and indivisible, a seamless web, a single inconceivable and transindividual process, in which there is no need to invent ways of linking language events and social upheavals or economic contradictions because on that level they were never separate from one another" (40).

Although he also argues, but "not [to be] understood as a wholesale endorsement of poststructuralism," that "only the dialectic provides a way of 'decentering' the subject concretely" (60), Jameson's insistence on "the reality of history" (82) suggests essential or preexisting, perhaps also unconscious, selves on which social identities are based regardless of the extent to which identities are obscured as constructed discursively by "transcoding," which Jameson defines as the strategic "invention of a set of terms" by which "levels of reality" are mediated, comprehended, and imbricated (40). As a result of this deep-seated structuralism, Jameson's new brand of Marxism made many second-wave feminists apprehensive and anxious about alliances between Marxism and psychoanalysis,[7] just as other feminists concerned with minoritarian struggles became skeptical and wary of Laclau and Mouffe's dismissal of classical Marxism's ontological privileging of class difference as vital to understanding historically socioeconomic, and thus social identity, transformations.[8]

The new sociopolitical, identitarian trends in critical theory, educational interest, and scholarship in general reflected and produced a phenomenon that was encapsulated in another popular catchword, "cultural anxiety." The title to Marjorie Garber's *Vested Interests: Cross-Dressing & Cultural Anxiety* (1992), a book that characterizes the first decade of the first wave of "gender studies," what we call the "engendered years," brilliantly captures the tenor of the times. Garber's *Vested Interests* was published by Routledge, a trade press that had quickly become the leader in the publication of books in the newfangled fields of cultural and gender studies. While Garber's book reflected the concerns of the era of its inception, the book that was to most influence the second decade of gender studies was Judith Butler's *Gender Trouble: Feminism and the Subversion of Identity* (1990), also published by Routledge (from the press' new series, "Thinking Gender"). In *Gender Trouble*, while implicitly dismissing the dialectic Jameson prescribes but remaining very much aware of the hegemony that is Laclau and Mouffe's focus, Butler argues that social identity is constituted only through performance (in speech, behavior, and style) and works to further a political agenda. Whether performed consciously or unconsciously, there is nothing essential or natural about sexual-gender identities. The categories "man" and "woman" are sociocultural constructs for which there is no original on which to base one's performance. These genderized, "performative" (Butler's adaptation of J. L. Austin's term) categories, affirmed or contested depending on the social actions repeated, are simply manifested ideas only inasmuch as people willfully or inadvertently subscribe to them.

The recent constructivist development in academic thought and research was paralleled in various popular media, perhaps most notably on film and television, a trajectory that began primarily with men cross-dressing as women (*Dressed to Kill* [1980], *Bosom Buddies* [TV 1980–82], *Tootsie* [1982], and *Mrs. Doubtfire* [1993]), then moved to include women cross-dressing or performing "male" (*Victor/Victoria* [1982], *Yentil* [1983], *Shakespeare In Love* [1998], and *Triumph of Love* [2001]), and ended up showing people not just cross-dressing as a means to an ends (such as for a job), but actually psychologically at odds with their designated gender identity, that which was socially-prescribed based on their sexual organs (*Silence of the Lambs* [1991], *The Crying Game* [1992], *Boys Don't Cry* [1999], and *All About My Mother* [1999]). In turn, and consistent with concerns of the time, the different approaches to "gender trouble" represented on film coincided with work done by literary-cultural critics on transvestism and gender difference on both the stage and street in early modern England. While Shakespeare's plays, particularly *As You Like It* and *Twelfth Night*, have received much attention, the play most consistently discussed in regard to both stage and street cross-dressings is Middleton and Dekker's *The Roaring Girl*.

Historiographically, *The Roaring Girl* is unlike any other extant play from the period in that it represents, by name, a living London personality as a character, and the playwrights' choice of "personality" for performance was a notorious gender-blurring icon of her own time, the infamous Mary Frith, also known as Moll Cutpurse. As suggested by Middleton and Dekker's character Sebastian's comparison of her to a religious icon ("I must now, / As men for fear, to a strange idol bow" [1.1.19]) and his reference to her notoriety ("a creature / So strange in quality, a whole city takes / Note of her name and person" [1.1.100–2]), Mary Frith/Moll Cutpurse achieved "celebrity" status in the seventeenth century, and into the twenty-first, because she publicly defied conventional dress codes and sumptuary laws by donning man's apparel and was, allegedly, an activist for this practice. According to a January 27, 1612 entry in *The Consistory of London Correction Book*, Mary/Moll refused, even after being officially condemned for her behavior, to refrain from unconventional acts of self-fashioning:

And further [Mary Frith] confesseth that since she was punished for the misdemeanors afore mentioned in Bridewell she was since vpon Christmas day at night taken in Powles Church with her peticoate tucked vp about her in the fashion of a man with a mans cloake on her to the great scandall of diuers persons who vnderstood the same & to the disgrace of all womanhood.[9]

Her public displays of transvestism, along with other "criminal" or transgressive practices, including cursing and thieving, cast her into the sociohistorical category, as Middleton and Dekker's character Laxton identifies her, of "Base rogue!" (2.1.271).

For many scholars today as well as for her contemporaries, Moll Cutpurse/Mary Frith's "celebrity" and/or "criminal" status make her an especially appealing historical figure of social deviance for a case study by which to proffer a theory of the early modern rogue, and of deviant identities and practices in general. In most of the early modernist scholarship discussed in the following pages, Moll/Mary exemplifies one or a combination of several recent postmodern formulations of subject positioning: 1) of what Laclau and Mouffe refer to as a "nodal point" (derived from Lacan's "*point de caption*," or "quilting point"),[10] which is an empty signifier capable of partially fixing and centering (like a Derridean "supplement") the substance of a range of floating signifiers (in this case, Mary Frith, Moll Cutpurse, rogue, roaring girl, transvestite, deviant, proto-feminist, queer, celebrity, etc.) by articulating them within a chain of equivalential identities among different elements that are seen as expressing a certain sameness (celebrity, queer, proto-feminist, deviant, transvestite, roaring girl, rogue, Moll Cutpurse, Mary Frith, and so on);[11] 2) of what Foucault calls "points of resistance," which are sources of antagonism, in such forms as dissenting or faltering subjects, within a society's strategic field of power relations (95–6); 3) of what Jameson considers subjective products of "transcoding," the sociohistorical processes by which one can "make connections among the seemingly disparate phenomena of social life" (in the case at hand, late twentieth-century feminism and early seventeenth-century criminality, among others); and 4) of what neo-Marxist Paul Smith identifies as an outcome of "the cerning of the subject," whereby "the contemporary intellectual abstraction of the 'subject' [such as of the historical persona Mary Frith] from the real conditions of its existence continues – and is perfectly consonant with – a Western philosophical heritage in which the 'subject' is construed as the unified and coherent bearer of consciousness," and which in effect works "to encircle" or "to enclose" the subject theoretically and rhetorically as a way of limiting the definition of the human agent in order to be able to call him/her the "subject" (xxx).

Throughout the course of the present analysis, we tango with each of these perspectives as they are reflected in the materialist and constructivist accounts of Mary Frith/Moll Cutpurse as a means by which to chassé into our own alternative theoretical understanding of deviant identity formations, in varied historical forms and different social contexts, as potentially "transversal agents." Our primary aim is to analyze the vast amount of criticism published on *The Roaring Girl* and Mary/Moll as an avenue through which to examine the preoccupations of recent literary-cultural criticism and the ideologies and methodologies that have fueled it in order to assess the impact of both Middleton and Dekker's *The Roaring Girl* and Moll Cutpurse/Mary Frith on contemporary academic discourse. Specifically, we are interested in the transhistorical phenomenon of what we call "Mary/Mollspace," particularly in how this "space" achieved such high status within the past twenty years and how it might influence the critical future. We will begin by explaining Mary/Mollspace in the process of delineating key components to transversal theory. Then we will review the different approaches to *The Roaring Girl* and Moll Cutpurse/Mary Frith, distinguishing them by ideology, methodology, and interpretation, as a way of moving into our own assessment of Mary/Mollspace as subject matter for a literary-cultural-critical-historiographical project.

Transversal initiatory, subjunctive space

From the perspective of scholars today, Mary Frith/Moll Cutpurse, competing with Christopher Marlowe, is possibly the most notorious cultural icon of England's early modern period. As with Marlowe, whose sexuality has also been a subject of debate since early modern times, the emphasis on gender differentiation in literary-cultural studies since the mid 1980s has contributed hugely to Mary/Moll's current iconic status. Resurrected in the postmodern period as a prototypical feminist and an early avatar of gay, lesbian, and/or transvestite power, Mary/Moll has become a vehicle for discourses on the blurring of sexual-gender roles and for feminist and queer politics generally. Moreover, she is emblematic of a criminal culture that has become increasingly important to scholars of England's early modern social history, especially as scholarship has become more aware of the period's complex sociocultural dynamics across class, religious, gender, and ethnic divisions.[12] Consequently, Mary/Moll provides a fertile landscape from which to harvest an understanding of the production and consumption of transhistorical "celebrityness," and in this case a celebrityness born out of social deviance. She testifies to the creation of a plurality of spaces that, like "Shakespace," moves discursively, and often transversally, conceptually, emotionally, and physically, across spacetime dimensions.

Shakespeare-influenced spaces, however conventional, alternative, or sometimes both, are what Donald Hedrick and Bryan Reynolds call "Shakespace,"[13] a term that accounts for these particular spaces and the time or speed at which they move from generation to generation and from era to era. For a variety of reasons, Shakespace is an especially strong exemplar of "transversal power," which refers to any force that induces change to a system, such as through altering an individual's subjectivity. Shakespace's unique functions as resistor, generator, and conductor of transversal power are resultant from phenomena that within and passing through Shakespace are the epochal forces and transformations wrought by a multiplicity of social, cultural, and economic influences: by early capitalism; by the new public entertainment industry in early modern England; by the interrogation of socially-prescribed gender roles; by aristocratic and legitimation crises; by the desacralization of absolutist sovereignty; by cross-cultural collisions and relativizations deriving from exploration and colonization; by the scientific revolution and its confounding of official knowledge; and – to follow this space into our own time – by the recursive force of the Western canonical tradition itself on Shakespeare's work.

The kind of iconic status *The Roaring Girl* has achieved is comparable to that of Shakespeare's *Romeo and Juliet*, whose title characters have become emblematic of a socially-prescribed form of "true love" – heterosexual, monogamous, and worth dying for – that has been officialized through spacetime in the image of its similarly iconicized author, William Shakespeare (see Chapter 6). The use of Shakespeare's play to promote a heteronormative conception of desire is an example of Shakespeare functioning as a "sociopolitical conductor," an ideo-logically-driven mechanism that works to create "subjective territory," an individual's combined conceptual and emotional scope. Defined succinctly, "Subjective territory is delineated by conceptual and emotional boundaries that are normally defined by the prevailing science, morality, and ideology. These boundaries bestow a spatiotemporal dimension, or common ground, on an aggregate of individuals or subjects, and work to ensure and monitor the cohesiveness of this social body" (Reynolds, "Devil's House" 146). In order to develop and sustain subjective territories, the "state machinery" of a society, an assembly of sociopolitical conductors (educational, familial, juridical, and religious structures), continually manufactures a dominant, "official culture."[14] As a result, "Official culture's sociopolitical conductors work to formulate and inculcate subjective territory with the appropriate culture-specific and identity-specific zones and localities, so that the subjectivity that substantiates the state machinery is shared, habitually experienced, and believed by each member of the populace to be natural and its very own" (Reynolds, "Devil's House" 147). Restated in quasi-Laclau-Mouffeian terms ("quasi" because of certain conceptual incompatibilities), subjective territory is informed by the differential relations of a sociopolitical environment's multiple subject positions and is itself comprised of multiple, spatio-temporally-specific nodal points that coordinate the subjective territory's multi-dimensional positionality relative to other subjective territories in the interest of promoting coalescence and fixity within both its own conceptual and emotional range and the greater sociopolitical field it occupies.

For Laclau and Mouffe, whose theory of subject formation comes predominantly from Lacanian psychoanalysis, the subject, like desire, is predicated on lack,[15] and it can only establish itself as a subjectivity through a common grounding and knotting together of equivalential identifications (subject positions with a recognizable and/or persuasive degree of sameness) into a singular nodal point that creates and endeavors to sustain a fully-achieved social identity. The nodal point, as Slavoj Žižek explains Lacan's formulation, "*as a word*, on the level of the signifier itself, unifies a given field, constitutes its identity: it is, so to speak, the word to which 'things' themselves refer to recognize themselves in their unity" (95–6). By extension, through linked equivalential identifications, society is also pursued by hegemonic forces. However, according to Laclau and Mouffe, such "suturing" together of seeming equivalences can never produce ultimately either a subject, an individual, or a society because there are always antagonisms from without that work hermeneutically to "overdetermine" (a Laclau-Mouffeian borrowing from Freud) the presence of some of the sutured social identities and social positions vis-à-vis others, causing either their condensation or displacement, and thereby transferring the focus onto another set of equivalencies so as to fixate another nodal point, and so on. Laclau and Mouffe's sense of stepping-stone subject formation is much like Derrida'a theory of *différance*, which accounts for the phenomenon that a signifier has significance given that it is a product of difference, but that its significance is simultaneously unstabilizable and always deferred because it can never achieve absolute meaning as a transcendental signified (see Chapter 1).[16]

Unlike Foucault, who sees antagonism as internal and "never in a position of exteriority" to power relations of a particular society (93–6), Laclau and Mouffe argue that "antagonisms are not *internal* but *external* to society; or rather, they constitute the limits of society, the latter's impossibility of fully constituting itself" (125). As Žižek eloquently summates:

> The thesis of Laclau and Mouffe that "Society doesn't exist," that the Social is always an inconsistent field structured around a constituted impossibility, traversed by a central "antagonism" – this thesis implies that every process of identification conferring on us a fixed socio-symbolic identity is ultimately doomed to fail. The function of ideological fantasy is to mask this inconsistency, the fact that "Society doesn't exist," and thus to compensate for us for the failed identification. (127)

But for Laclau and Mouffe it is not only society that is negated, but also the identifiable human agents who might have sociopolitical power should it be possible for them to be members of a society. Laclau expresses this concept succinctly:

> The question of *who* or *what* transform social relations is not pertinent. It's not a question of "someone" or "something" producing an effect of transformation or articulation, as if its identity was somehow previous to this effect. [...] It is because the lack is constitutive that the production of an effect constructs the identity of the agent generating it. (210–11)

Articulated in the space of "everyday life," should you want a mechanic to work on your car, you might as well solicit anyone to assist you, because it is impossible for you to know who is or is not a mechanic until a person demonstrates his/her mechanical skills, and even then, we might ask, "How skilled must he/she be to be a mechanic, and who is qualified to make this evaluation?" Of course, we might respond to such queries with, "Would seeking a state-certified mechanic at a state-licensed automobile repair shop increase the odds of her being skilled enough?" Empirically, if what appeared to need fixing appears to have been fixed by the mechanic, then chances are it was. Within the relevant, local interpretive community, agency was identified and meaning was achieved, as predicted by the pre-established parameters of the quest for car repair. Transposing this example to one more in line with Laclau and Mouffe's political agenda, we might ask, "Doesn't the revolutionary potential of individuals, perhaps measured by past expressions of dissidence, matter when pursuing alliances for the revolution?"

Alternatively, transversal theory and methodology are all about potential. They work discursively to empower social identities and groups recognizably striving – conceptually, emotionally, and/or physically – to transcend their subjective territories. As stated, according to transversalism, the idea of totalized or absolute subjectivities, identities, societies, or states is an obvious impossibility insofar as movement and change are fundamental to existence. Transversal theory's subjective territory, unlike the always already lacking and failed subject for Laclau and Mouffe, is not an ahistorical phenomenon continuously manifested, expanded, and discombobulated through just antagonism, but a space socially, culturally, and politically generated and maintained through antagonisms as well as other conceptualizations and experiences, including evolution, curiosity, imagination, empathy, desire, and love. As described above, subjective territory is a conceptually- and emotionally-constrained, multi-dimensional space through which humans with social identities and varying degrees and kinds of agency navigate their consciousness and presence within specific social, cultural, and historical parameters of spacetime. Subjective territory is an unfixed, permeable, and processual space that thrives on its interactive relationship with "transversal territory."

In contrast with, and in opposition to, state-orchestrated subjective territory is transversal territory, the non-subjectified space of one's conceptuality and emotionality. According to transversal theory, "transversal movement," the conscious and/or subconscious breach and transcendence of one's subjective territory into either the subjective territory of an other(s) or a non-delineated alternative(s), threatens the stability of the state machinery and its regulating, official culture. Transversal movements are propelled by transversal power, a fluid and discursive phenomenon defined as any force – physical or ideational, friendly or antagonistic – that inspires emotional and conceptual deviations from the established "norms" for an individual or group. Transversal movements indicate the emergence and inhabitance of transversal territory, a chaotic, boundless, challenging, and transformative space through which people traverse when they violate the boundaries of their prescribed subjective addresses. Transversal territory is a conceptual-emotional space people inhabit transitionally and temporally when subverting the hierarchicalizing and homogenizing assemblages of a governing organizational structure. To reside permanently in transversal territory, rather than pass through it, would be to subsist without any kind of cognitive stability or control. In effect, transversal territory threatens "official territory," the ruling ideology, propriety, and authority that provide the grounding and infrastructure of a society, by offering a space through which to transcend it.

Passing through Shakespace

Hedrick and Reynolds' term "Shakespace," which accounts for the "diverse and numerous" past, present, and future workings of Shakespeare through "countless commercial, political, social, and cultural spaces" (8), is a helpful term for analyzing the various literary, cinematic, official, and unofficial constructions of Shakespeare since it accounts for the elongated history of the official, "subjunctive," and transversal uses of Shakespeare. As with other icons whose work and "affective presence" – a subject/object's "combined material, symbolic, and imaginary existence" that "influences the circulation of social power" (Reynolds, *Becoming Criminal* 6) – have spawned transversal spaces, such as Jesus, Marx, Freud, and Osama bin Laden, the dissemination across spacetime of Shakespeare and his plays, propagandized as icons of official culture, has "encourage[d] alternative opportunities for thought, expression, and development," just as it has, alternatively, "promote[d] various organizational social structures that are discriminatory, hierarchical, or repressive" (Hedrick and Reynolds 9).

As much as Shakespeare has been constructed and utilized to impose limitations on social behavior, his characters, like Middleton and Dekker's title character in *The Roaring Girl*, often engage in transversal movements, defying or surpassing the boundaries of their prescribed subjective addresses, opening themselves, as it were, to subjective awareness outside the self or selves. In Shakespeare's plays, ruptures in subjective territory frequently occur through the transversal, transformative process of desire, a productive process of "becoming-other" defined against the desire constructed and regulated by and for state power in the interest of maintaining official culture:

> Transversality, on the other hand, produces and expresses desire in the dynamic form of what Gilles Deleuze and Félix Guattari term "becoming." Becoming is a desiring process by which all things (energies, ideas, people, societies) change into something different from what they are; and if those things were, before their becoming, identified, standardized or normalized by some dominant force (state law, Church credo, official language), which is almost inevitably the case, then any change whatsoever is, in fact, a becoming-other. The metamorphosis of becoming-other-social-identities

trespasses, confuses, and moves beyond the concepts of negation, essentialism, normality, constancy, homogeneity, and eternality, that are fundamental to subjective territory. (Reynolds, "Devil's House" 150–1)

A quick survey of Shakespeare's plays would reveal desire as an indispensable and transformative process undertaken by many of the main characters to achieve a new social identity, becoming-an-other within, without, and/or against an official culture: Macbeth, conforming to the prevailing ideology of his subjective territory, follows his ambitious desire to become a king; Desdemona's passion for and ensuing marriage to Othello in spite of racial prejudices challenges the boundaries of a dominant discourse on desire; the exiled Prospero, in enacting his revenge upon King Alonso and his brother Antonio, the current Duke of Milan, seeks to disrupt state power from afar; and Iago, operating from within Venice's official culture, plans to destroy Othello, thereby overthrowing state power. From different angles of readership (and we cannot emphasize enough the flexibility with which the examples can be read, even contradictorily), these Shakespearean examples illustrate processes by which identities are transformed and thus revealed to be non-essential, socially-constructed designations: Macbeth moves swiftly up, then down, the ranks of feudal power; Desdemona becomes "our great captain's captain" (2.1.74), a whore, a victim, and then a martyr; Prospero reclaims his lost dukehood and foregoes his magic power, and with it his reign over Sycorax's island; and Iago is transformed from the state-loved, seemingly "Honest Iago" (1.3.295) to tortured captive of the Venetian state.[17] Like these Shakespearean characters, Middleton and Dekker's Moll Cutpurse experiences ongoing processes of identity formation, performance, and transformation that extend far beyond the play into and through the various discourses in which Moll Cutpurse/ Mary Frith journeys transhistorically.

The Roaring Girl and Moll/Mary emerged out of roughly the same historical context – early modern England, specifically London, Southwark, and the theater community – and traveled both diachronically and synchronically in a similar, although lower-profile, scholarly discourse as Shakespeare-as-literary-cultural-icon has since the mid 1980s. For this reason, we imagine Mary Frith/Moll Cutpurse, a duel historical persona, occupying a comparable discursive space, the aforementioned Mary/Mollspace, that embodies and metamorphs a variety of identity formulations, from heterosexually married woman to rogue to drag king to a representation of the Madonna–whore dichotomy. We see the case of Mary/Moll as being exemplary of a sociohistorical phenomenon that is also a historiographical dilemma: What to do when a real-life historical figure is multifariously re-envisioned and appropriated transhistorically in diverse narrative modes, including biography, fiction, performance, and critical discourse? Using the case of Mary Frith/Moll Cutpurse, we will demonstrate the value of transversal poetics' critical methodology, known as the "investigative-expansive mode of analysis" (i.e. mode), a critical approach that first breaks down the subject matter under investigation into variables and then partitions and examines them in relation to other influences, both abstract and empirical, beyond the immediate vicinity.[18] A chief objective of this expansive methodology is to contextualize historically, ideologically, and critically both the subject matter and the analysis itself within local and greater milieu. A mobile approach, the i.e. mode continually reparameterizes in response to the unexpected emergence of glitches and new information, and resists anything resembling predetermination or circumscription. The employment of the i.e. mode allows us to explore fugitively the ways in which the case of Mary/Moll exemplifies "subjunctive space," the hypothetical space of both "as if" and "what if," which is our overarching purpose in the upcoming analysis.[19]

Because of its openness and uncertainty, subjunctive space is like "transversal territory," whose indeterminate mappings can occupy, transgress, and expand – to borrow Raymond Williams' terminology – the "residual," "dominant," and "emergent" aspects of a culture. In this respect, subjunctive space is a subset of transversal territory or, more accurately, it is an inbetween space operative between subjective and transversal territories in which the subject necessarily retains agency and can self-consciously hypothesize scenarios and experiences, thereby self-activating her/his own transversal movements. Yet because of its telegraphed parameters of potentiality, predicated on or manifested by the "suggestive information" identified within the subject matter under investigation (whether it is means of time management, schisms in religious teachings, or a theatrical text, to name some), subjunctive space can also spur resistance to transversal territory insofar as it allows for and includes all potentialities. Put differently, the anticipated, contingent and/or possible "ifs" that can work to empower the subject through transversal movement can also stabilize, empower, and/or disempower the subject by further subjectifying her/him within her/his prescribed subjective territory.

Supporting Derrida's answer of *différance* and supplementarity to the problem of logocentrism and hermeneutics focused on ontology, subjunctive space also provides a way out of both the circumscribing Husserlian focus on intentions and their objects (of both readers and authors, for instance) and the Geneva School's, and later reader-response critics', enhanced focus on the consciousness of individuals (of both readers and authors) that has characterized phenomenological literary-cultural criticism and has informed much neo-conservative humanist criticism since the mid 1980s, such as, in English Renaissance studies, that of Richard Levin and Brian Vickers.[20] Subjunctive space accepts all de-centering, disseminating, deconstructive machinations as a means by which, in Paul Smith's terms, the "subject/individual" ("the human entity to whom qualities of being a 'subject' or an 'individual' are commonly assigned") can become discerned or discern his/herself (xxxv). Subjunctivity allows the subject to become at least provisionally uncontained and unconstrained so as to point in certain directions and to certain places, to past, present, and future elsewheres indicated and imagined by the projections of "as if" and "what if." Moreover, and contrary to the Derridean perspective, whereas the subject may be vulnerable to deconstructionist procedures because the subject is already a negotiated construct within a fluctuating and often struggling subjective territory, according to transversal theory, the individual always negotiating – consciously and/or not – his/her subjectification and social identity in relation to transversal power cannot be completely deconstructed and/or contained, regardless of whether the power is willfully generated by the individual and/or manufactured through engagement with the social factory of which he/she is a product. When the deconstructionists' rhetorical dust settles discursively or continues discursively to cloud the air, no less than a phenomenon remains: a human idiosyncratically experiencing a field of consciousness and possessing degrees of agency and potential peers on.

In subjunctive space, then, the individual is neither necessarily forever pulling the rug out from under her/his fugitive feet as she/he tries to escape the logocentrism infrastructurally reinforcing her/his subjective territory, the sociopolitical conductors inscribing and maintaining that territory, and the state machinery in which those conductors operate; nor is she/he ineluctably and desperately striving to concretize her/his consciousness in cahoots with the intentions of an author, group of authors, and ruling authorities or the state, thereby, in theory, achieving self-value through the actualization of an aesthetic object in the living world. Instead, subjunctive space opens up opportunities for coordinating one's own subjective territory vis-à-vis competing ideologies and objects/subjects to the extent that one can imagine alternative conceptual, emotional, and physical coordinates. In the case at hand, the discourse and those exposed to the discourse of Mary Frith/ Moll Cutpurse, including the writers and readers of this essay, are the subjects and subject matter under investigation. Venturing now into a subjunctive space while moving backwards and forwards investigative-expansively, we journey to and through historiographical coordinates and discourses on *The Roaring Girl*, associated celebrity spheres – from the Virgin Mary to Christopher Marlowe to the modern Madonna – and the interrelated subfluxation of Mary/Mollspace.

Still roaring after all these years

Mary/Moll has occupied various spaces since the emergence of her publicized criminal persona in the seventeenth-century,[21] when her notoriety as a cross-dressing cutpurse manifested itself in her appearance as the subject of popular literary and theatrical fare. During her own lifetime, she appeared in multifarious textual guises, from criminal in court documents, to subversive recalcitrant at a public penance immortalized in John Chamberlain's 1612 letter, to onstage performer in the afterpiece advertised in the Epilogue to Middleton and Dekker's play.[22] Her social deviance and appealing personality made her a newsworthy subject for myth-making discourse. Shortly after her death in 1659, she was reborn as a royalist cross-dressing Robin-hooder who eventually seeks redemption in *The Life and Death of Mrs. Mary Frith, Commonly called Mal Cutpurse* (1662), a three-part self-declared biography that includes the "real" Mary Frith's supposed diary.[23] After her rebirth during the Restoration, Mary/Moll remained relatively quiet on bookshelf coffins until the twentieth century, when she was revived, first in an effort initiated in 1927 by T. S. Eliot to rescue and/or redeem Middleton's playwriting skills from critical dismissal by focusing on the "realness" of *The Roaring Girl*'s title character: "Middleton's comedy deserves to be remembered chiefly by its real – perpetually real – and human figure of Moll" (99).[24] While Eliot's essentialist mission to salvage Middleton and Dekker's Moll Cutpurse – for no other reason than because she is a "real and unique human being" (89) who "remains a type of the sort of woman who has renounced all happiness for herself and who lives only for a principle" (96) – is often cited today in critical discussions of the play, *The Roaring Girl* went dormant for another forty years until the emergence of the "sexual revolution" and first-wave feminism in the West. Beginning in the 1970s, Mary/Moll began to be mobilized via productions of Middleton and Dekker's play to address feminist issues.[25] Academics followed suit beginning in 1984 with the publication of Mary Beth Rose's "Women in Men's Clothing: Apparel and Social Stability in *The Roaring Girl*."[26] From 1984 to 2003, rescuing and/or redeeming Mary/Moll, either from literary and/or historical obscurity, from her representation in *The Roaring Girl*, or from mythologized readings of her based on "fact" or fiction, has become the *modus operandi* of critical considerations of her deviance and dissident potential in both the public sector of early modern England and the institutionalized realm of postmodern scholarship. In recent years, Moll/Mary has been salvaged and revitalized in numerous shapes and sizes, ranging from proto-feminist heroine, dissident disrupter of dominant gender and sexual codes, and less-than-transgressive transvestite, becoming a contested site/sight for feminist, queer, new historicist, and/or materialist discourses.[27] As our brief examination of the more recent critical constructions of Mary/Moll will show, redeeming the "real" Mary and/or the theatricalized Moll has generated multiple, and sometimes overlapping, Mary/Mollspaces in which she is reconfigured, redisplayed, and re-remembered as an academic icon of dominant critical discourses and as a spearhead for various, primarily liberal/leftist, political agendas.

As Eliot predicted, Middleton and Dekker's *The Roaring Girl* has been "remembered chiefly" for its recreation of the "real" (99). The majority of critical discussions of Mary Frith/Moll Cutpurse's potential for disruption of the dominant order offstage have focused on Moll as represented on the stage of the Fortune Theatre in the original 1611 production of the play.[28] With scant "hard" evidence of the historical Moll's life, critics have turned their attention to the seemingly more "stable" theatrical text. The play itself is atypical of the seventeenth-century stage, not only because of its unconventional city comedy conclusion,[29] but because of its real-life, then still-living subject. As Katherine Eisaman Maus notes, the Moll of *The Roaring Girl* is "the first positively identifiable living person to be translated into a quasi-fictional dramatic realm" (1371). Maus considers Mary Frith's social status as the cause for the exception to early modern England's censorship rules, which "prohibited theatrical depiction of politically powerful, socially prominent people" (1371). While Mary's non-"socially prominent" class status may have granted the playwrights latitude when breaching contemporary censorship laws, Mary Frith the icon nonetheless functioned as a space through which Middleton and Dekker could produce the previously unpresentable. Recasting the "real" in the representational, Middleton and Dekker generated the perhaps first in a plurality of Mary/Mollspaces that have continued to emerge and expand into the twenty-first century. Their version of Moll, while debatedly transgressive in relation to their official culture's gender and/or sexual codes, subverted state-imposed dramatic conventions. By exposing the ban on blurring the fictional with the living-real to be blurrable itself, Middleton and Dekker opened doorways, later domino-effected by Eliot's campaign, not only for future theatrical rule-breaking, but for modernist and postmodernist challenges to both social identities and to sociopolitical conductors informing and performing identity formations.

Crossing thresholds into subjunctive space, whether championing the theatrical or "real" Moll, modern critics have mobilized Mary/Moll to investigate structuralist and poststructuralist configurations of, in transversal terms, the sociopolitical conductors that inscribe subjective territories and prescribe social identities. The majority of recent criticism on Mary/Moll can be succinctly categorized – taking into consideration variances and overlappings – according to four major "camps," three of which have revived Mary/Moll as an exemplar of the performability of gender and/or sexual subject-positions: 1) feminist/queer critics who, focusing on *The Roaring Girl*, identify the theatricalized Moll as a progressive transgressor of gender, sexual and/or class systems (Rose, Comensoli, Miller, Dollimore, Howard, Orgel, Mikalachki, Kermode, Rustici, Heller, and Reynolds); 2) feminist/queer critics that, seeking to demystify the Moll Cutpurse of Middleton and Dekker's recreation, position the "real" Moll, in contrast to the play's representation of her, as radically oppositional to dominant views of sexual, gender, and class differentiations both in early modern England and today (Krantz, and Baston); 3) feminist/queer/materialist/(new) historicist critics that, focusing on *The Roaring Girl*, present the scripted Moll as a performed affirmation of a patriarchal order who has been misread as radically progressive (Mulholland, Garber, Jacobs, Krantz, Baston, Forman, West); and 4) historiographers that focus on the 1662 biography to contest or support previous politicized readings of the "real" and theatricalized Moll Cutpurse (Nakayama, Todd and Spearing, and Ungerer).[30] In the course of these diverse readings, multiple Molls have been constructed and deconstructed, suggesting the very subjectivity of "reading" identity, the potential for re-forming identities, and the mutability of the seemingly static page.

Since her 1984 "rescue" by Rose, Mary/Moll has been discursively reconfigured within the four camps, sometimes simultaneously, as: 1) an early modern exemplar of women's lib (Rose, Comensoli, Miller, Baston, and Kermode); 2) a conduit for destabilizing dichotomous gender and/or sexual structures (Dollimore, Howard, Orgel, Nakayama, Todd and Spearing, Mikalachki, and Krantz); 3) a cant-talking, class structure breacher in breeches (Orgel, Mikalachki, and Reynolds); 4) a homoerotic spectacle for mass male consumption (Garber, and Orgel); 5) a celebration of homoerotic desire (Howard, and Heller); 6) an emblem for the theater itself (Heller); 7) a conflated figure of the cant-pamphlet producing Dekker (Mikalachki); 8) a "compensatory fiction" for the commodification of identity via capitalism (Forman); 9) a commodified product of the then-emerging market culture (Ungerer, and Forman); 10) a celebratory figure of the subversive potential in smoking the herb of the "real" and representational roaring girl's choice (Rustici); and 11) a mystified, misread, "transvestiting," non-transgressing affirmation of a patriarchal, dominant order (Mulholland, Garber, Jacobs, Ungerer, Maus, and West).

In all of these formulations, manipulations, and interpretations, Mary/Moll has occupied discursive, cross-pollinated spaces that, moving transhistorically from the seventeenth- to the twentieth- and into the twenty-first century, have sought to redefine, reclaim, and redeem what Moll Cutpurse has been and is becoming over 325 years after Mary Frith was. Apparently deeply invested in early modern England as a foundation for the postmodern present,[31] these critics have generated contradictory and sometimes self-competing Mary/Mollspaces that map out subjective territories from the "then" and project them onto the "now." While Moll/Mary has been jolted forwards, backwards, and across spacetime in her projected (or not) political radicalness, she has retained her image, archivally emplotted, as a public persona in the criminal celebrity sphere of England's early modern period. Despite Eliot's insistence that the roaring girl "deserves" to be remembered as a "human figure" (99), Middleton and Dekker's version of Moll is generally read less as "human" than as a sociohistoricalizing, contextualizable, still-living museum piece of dissidence and/or subservience. She has become a sign-object – the biunivocal (two-into-one) conceptual and emotional expression of the different signifiers and signifieds associated with her – that has been fetishized, like the Virgin Mary and the modern popstar Madonna, as both commodity and political cathexis. Through fetishization, the sign-object Moll/Mary has achieved affective presence; its amalgamated material, ideational, symbolic, and wish-fulfilled subsistence and becomings have produced, reproduced, and delineated Mary/Mollspace.

Whether spun by recent scholarship as a proto-feminist icon, as a reformed version of the "real" transgressor (Moll's offstage, historical counterpart), or as a mythologized commodity born of a market culture manufacturing commodified criminality, Moll/Mary has been rendered into a product and symbolic figure of early modern England's emergent capitalist system, a figure of either exploitation by, from within, or of systems we would define as organized clusters of socio-political conductors. Under, against, and/or in support of the maintenance of the conductors' corresponding official and subjective territories, Moll/Mary was generated and generates alternative selves, thereby initiating and perpetuating a becomings-Mary/Mollspace with an enduring affective presence whose substances, traces, and phantasmagoria effectuate transversal movements into both navigatable and unchartable territories across spacetime.

Subjunctive mappings, critical coordinates

Moll/Mary has become a spacetime-traveling iconicized celebrity that reveals an inclination to find a precedent for contemporary methods of dissent. This is the case especially in interpretations that position Mary/Moll as a "progressive" subvertor of a dominant order. Employed by feminist, queer, new historicist, and materialist critics, as we have seen, Mary/Moll, like Shakespeare, has been mapped across spatial, temporal, emotional, conceptual, and sociocultural boundaries; thus, she has been imagined by some English Renaissance scholars of the postmodern age (including Rose, Garber, Orgel, Howard, Spearing and Todd, Kermode, and so on) as a transhistorical transgressor and/or affirmer of official culture. By projecting Mary/Moll into simultaneous pasts, presents, and futures, these critics have rejected conventionalized spatiotemporal demarcations. However unproblematic or not, the projections of Mary/Moll has led some down the critically disavowed path of universalization plaguing, in Margreta de Grazia's estimation, new historicist literary-cultural criticism.

In reference to the "early modern" labeling of proto-capitalist England, de Grazia has noted the predilection among new historicists to analyze the present by its relation to a period of history conceived of as a precursor to the now. De Grazia rejects the label "early modern" as a universalizing technique counter to the new historicist project: "There is a way in which seeing the Renaissance as the Early-Now commits itself to the very universalizing tendency that historicizing set out to avoid in the first place. As if *the* relevant history were a prior version of what we already are and live" ("Ideology of Superfluous Things" 21). Craig Dionne reiterates de Grazia's criticism, noting that "the self-referential historicist strategy" he identifies as characteristic of new historicism "can potentially negate the difference of the past by rendering early modern social activities as a kind of allegory of the present" (38). Deborah Jacobs makes a similar claim in reference to what we would term "feminist early modernist" readings of Middleton and Dekker's recreation of Moll Cutpurse. Arguing against "novelized" reading which "renders its own motives and politics invisible and remakes in the reader's own image" (75), Jacobs critiques new historicists' "willing[ness] to historicize context but not an individual subject's consciousness" and materialist "feminist discourse that is willing to rigorously historicize the material conditions of women's existence but still retains a transhistorical resistant 'woman' and is, furthermore, determined to find 'her' in other cultures" (79–80). Like the precursory "early modern" in de Grazia's analysis, and the renderings of past social practices into an allegorical present noted by Dionne, the transhistoricalized, proto-feminist Moll Jacobs critiques runs the risk of universalizing and thereby erasing social, cultural, and historical differences by identifying potentially resistant women then as prototypical of "Early-Now."

While Jacobs focuses her critique on interpretations of Moll as proto-feminist, citing Rose and Howard as prototypical of the "novelized" reading she dismisses, her assessment of this transhistoricializing approach can also be applied to critical disavowals of Mary/Moll's transgressive potential. By judging the early modern "real" criminal/English theatrical character according to contemporary evaluations of what constitutes dissidence, Mary/Moll has emerged in some circles as not transgressive enough, as incapable of transcending the official territory that, in her own time, inscribed Mary Frith as a criminal element in a hegemonic order. Through processes that historicize the present in their historicism of the past, Mary/Moll has been transcribed into, as Eliot remarked in 1930, a figure capable of being "perpetually real" (99), and, as we have shown, an icon capable of being subjunctively affective.

The transhistoricalizing processes through which Mary/Moll has been revamped disclose more than the subjectiveness of the interpretative act. As Jacobs argues, Mary/Moll has been reconfigured through a series of "novelized" readings which "renders its own motives and politics invisible and remakes in the reader's own image" (75). Visibly written from the position of politicized author motivated by what she assesses is critical misrepresentation, Jacobs' devaluation of the "novelizing" formations of Mary/Moll suggests that objective interpretation is an achievable goal, a goal that her own article throws into doubt. Selecting and dissecting evidence in support of an argument is a process of manipulating the material to support "the reader's own image" as the authoritative figure on the given material. Assessing the subjective vacuum of the interpretative act is not the issue here; rather, our focus is on the critical spaces Mary/Moll has occupied and transgressed in response to discourses developed to address subjective awareness and the positionality of the subject within its subjective territory. Like the Virgin Mary, Mary Frith/Moll Cutpurse has been appropriated and mobilized by forces in- and de-scribing residual (from early modern English to the present), dominant (contemporary official cultures), and emergent (critical future) landscapes; it is her very de-centeredness, her lack of fixity and a consistent referent in the historiographical record, that allows for continuous reformations of her as both a formative and performative figure of sociopolitical structures. Immortalized in her own time primarily in fictional modes (theatrical and "biographical") and in second-person narrative accounts of her public behavior, and then revived in the now for her behavior and/or representation in the then, Mary/Moll has become a paradoxical figure incapable of containment by a single discourse or narrative mode. Moll/Mary is mostly known today as a mythologized figure of the imagination, and is thus open to reinscriptive imaginings of her social deviance and transversal agency and power.

Mary/Moll, fetishized or otherwise, presently occupies what we call the subjunctive space of "as if/what if." In the debate since the 1980s over her potential dissidence, as we have seen, scholars have projected her into a realm of possibilities by considering her "as if" she were a proto-feminist, a homoerotic spectacle, a conservative commodity, and so on, inscribing her within the framework of "what ifs." Functioning in the subjunctive realm of possibilities of future-, present-, past-, and absent-space, Mary/Moll has become and invigorates multifarious hypotheticals that allow for imaginings of her potential as a disruptive force both diachronically and synchronically.

Novelized Moll: venturing further into "iffyness"

Continuing to move investigative-expansively into subjunctive space, we want to consider yet another rendering of Mary/Moll, something less in the recent tradition of academic discourse, but rather in the mode of biography. As has been shown, the "as if/what if" processing of Mary/Moll has primarily operated in the realm of academic configurations of Mary/Mollspace. However, Ellen Galford's novel *Moll Cutpurse: Her True History* (1985), self-consciously-framed as a "novelized" reading of Mary/Moll, projects Moll into the realm of "subjunctivity" by asking "what if" Moll Cutpurse were a lesbian, Robin-hooding proto-feminist, and then by telling her "real" story "as if" Moll "really" was. The novel's title advertises the work as "True History" ("as if" History needs to be qualified), but the "Historical Note" at the conclusion describes the text as a fictional improvement upon the factual sources on which the book is based: "Some of the episodes in this story are derived from these sources: the others may be as close – or closer – to the truth" (221). The novel positions this Moll as a version "close – or closer – to the truth," and the book functions as a "rescue" of the "real" Moll from fallacious textual renderings. As evident from the multivariate Molls redemptively constructed in the discourses previously discussed, "the truth" imagined by Galford is, despite the author's claim of a less distant relation to authenticity, only one of multiple potential truths or "what ifs."

In the Moll narrative as told by Galford, the "if" of Moll's sexual-gender-social identity is projected onto Bridget, the former lover of Galford's Moll. Bridget, an apothecary Moll went to see to inquire about a sex change – " 'Turn me into a man,' was all she said" (14) – re-remembers and narrates the now-deceased Moll's life for the reader. As established in the narrative, Moll was closeted about their relationship at Bridget's request, but now that Moll has passed away, Bridget is coming out to "rescue" Moll from male-authored renderings of Moll Cutpurse: "Now, so many years later, I feel the time has come to yield up that secret. Or all that will be left to keep alive Moll's memory will be the fabrications of *men*" (12). Those fabrications include *The Roaring Girl*, a play which did not displease Moll, since "those who wrote it were her old-boozing companions in any case" (12), and the *Life and Death of Mrs. Mary Frith, Commonly Called Mal Cutpurse*, written by a "dull scribbler who never knew her [but] has taken it upon himself to tell her story" (12). The impetus for the story is Bridget's coming out, and Moll's life history is conflated with the autobiography of Bridget, the novel's narrator with whom the reader is led to identify with: "So it falls to me, because I knew her best and loved her most, to tell Roaring Moll's true story, and my own with it" (12).

In Galford's appropriation of the romance novella form, Mary/Moll functions as a space not only through which lesbian relationships are normalized for a non-heterosexualized target audience, but through which autobiography – a process in which one can construct a self or selves for a reading public – can be scripted. Through the retelling of Moll's "True History," Bridget can author her own sociosexual identity; however, Bridget's personal inscription is not self-fashioned, bur rather conditioned by her sociosexual referent, the symbolic figure of Mary/ Moll. Bridget's self-formulation is dependent upon her reclaiming of "Moll's memory" (12), and, as such, she is defining her self in relation to a fetishized other. Furthermore, the primary motivation for Bridget's identity-construction is not a desire to produce a self unrestrained by her own subjective territory, but to recuperate Moll from male-orchestrated commodifying processes which have challenged Moll's value as a sign-object for non-heteronormative sexual practices and breaches of gender-power structures: as Bridget states, if Moll is not reauthored by Bridget, Moll will remain inscribed by "the fabrications of *men*" (12). Bridget's transcription of Moll Cutpurse's "True History" pro-duces a Mary/Mollspace consciously seeking to counter a dominant patriarchal structure by appropriating and fetishizing an icon misrep-resented by members of that order ("*men*"). The framework of the novel demonstrates its own investment in and mobilization of Mary/Moll as a phenomenon of the then with affective power in the now. Projecting Bridget and Moll into the subjunctive space of possibilities is framed as a personal enterprise affectively operating against sociopolitical conduc-tors of present-space subjective territory, for if Moll is not retranscribed, Bridget (the reader/the author) cannot tell her "own story" (12). In this subjunctive journey into Mary/Mollspace, Bridget's (and the reader's/ the author's) identity-formation is dependent upon the hypothetical of Moll Cutpurse's "iffy" sociosexual identity.

Rematerialized Moll/Mary: becomings-Marlovian

Rather than wrapping up this essay with the expected clever conclu-sion, a finale that subtly reenacts the gist of the arguments as it directs the readers intriguingly to grander implications off the stage of the page, we considered returning to *The Roaring Girl* by offering our own close-reading of the representation of Middleton and Dekker's Moll Cutpurse. Having debated whether or not to throw our two-cents into the already overflowing critical cup, we opted for another, uncharted strategy more in keeping with our own transversal enterprise. Before embarking on this voyage, let us clarify our reasoning for our readers, whom we imagine are wondering why we have not also chosen to "redeem," denounce, heroize, and/or fetishize Mary/Moll. Having mapped out previous literary-cultural-historical criticism of *The Roaring Girl*, we agreed that we have little new to offer in terms of readings of Moll's representation in the play. Instead, by manufacturing our own Mary/Mollspace out of various "historicalities" (material, theoretical, symbolic, and spectral vestiges), we prefer to propel the Mary/Moll of fact and/or fiction into the comparably subjunctified realm of Marlowespace. By merging these emotional, conceptual, and historical spaces, we hope to further illustrate the mechanics and potentialities of subjunctive ventures, and to briefly explore and point towards further expansions of Mary/Mollspace that move beyond the constraints and tautology of the established critical discourse on Mary/Moll and *The Roaring Girl* so that new critical, pedagogical, and perform-ance frontiers can be discovered. In our now-embryonic "Mary/Marlowespace," we ask "what if" Mary/Moll was Christopher Marlowe and/or imagine *The Roaring Girl* "as if" it were penned by Marlowe. By thrusting Mary/Moll and Marlowe into a joint subjunctive sphere, we posit one of the plethora of possibilities on the horizon of critical considerations and mobilizations of Mary/Moll, Marlowe, and the con-tinually reassessed and expanding early modern English canon.

In our fabrication to follow, Mary/Moll is metamorphed with Marlowe, the "roaring" Moll's possibly leading rival historical figure of potential dissidence in early modern English studies. Endowing Marlowe with the symbolic value of Mary/Moll, and Mary/Moll with the literary-historical value of Marlowe, we produce "Mollowe," a cross-dressing, playwriting, cape-crusader who, operating much like the malcontented heroes of Marlowe's drama,[32] is capable of transgression accompanied by disruption that we characterize as transversal move-ment. Appropriating Act Three, Scene One of *The Roaring Girl*, the scene most frequently cited in scholarship on the representation of Moll Cutpurse in the play,[33] we welcome our readers to join us in this brief, concluding excursion into Mary/Marlowespace.

> *Scene: Night, Gray's Inn Fields. Enter Saxton*
> SAXTON: Who's there?
> *Enter Mollowe, dressed like a man*
> SAXTON: Nay, answer me. Stand and unfold yourself.
> MOLLOWE: [Aside.] Oh, here's my frizzied gentleman. Like Faustus in prestige, but with Wagner's boyish looks. Little does he know he is to be robbed in the Inn. How his eye is like a Robin's, but he lacks his clownish Dick. [To Saxton.] Come sir, the readiness is all?

SAXTON: Ho, sir, ready for what?

MOLLOWE: Ingrammercy, do you ask that now, sir? You, who have conjured me?

SAXTON: [Aside.] "Conjure me"? Egads, is it you, Kit? [To Mollowe.] You seem some familiar. Have we known each other?

MOLLOWE: Thou art too ugly to attend on me. [Mollowe starts to go.]

SAXTON: Who's this? Kit? Not Kit, but Honest Moll? Stay, I charge thee.

MOLLOWE: Not since Deptford have I been thus hailed.

SAXTON: Hark, what word from yonder cloak doth break? "Deptford"? Are thee not Kit, schooled in night to divide the day? [Reaching to touch Mollowe.] A phantom from fashion of himself? I did love you once!

MOLLOWE: Nay, I said "Stepford!"

SAXTON: Faites excuse! I mistook you for a university wit I once well knew. Alas, Moll, of most excellent fancy.

MOLLOWE: Purblind? You're an old wanton in your eyes, I see that. [Removing cloak.]

SAXTON: No, not here. We shall be espied.

MOLLOWE: [Drawing sword.] Aye, there's the rub. [Showing money.] Here's the gold with which you hired your hackney. Racking hard, your bones will feel this. Ten angels of mine own I've put to thine. You lay the odds on the weaker side.

SAXTON: Hold, Moll! Mistress Mary. Strike a woman? I'm afeard you make a wanton of me.

MOLLOWE: [Stabbing at Saxton] A rat! Dead for a ducat, dead.

SAXTON: [Backing away.] Call me not "rat"? My rodent name from Wittenberg. [Draws sword.] By heaven, figured like the poet that's dead! Haste me to know it, that I with wings as swift as thoughts of love may resist your revenge. [They fight. Mollowe wounds Saxton.] A hit, a very palatable hit.

A cage descends from above

MOLLOWE: You're fat, and scant of breath.

SAXTON: [Wounded, crawling as Mollowe approaches.] Kit, forgive me my transgressions, I knew not what I did when I stabbed you over the right eye with a depth of two inches and the width of one.

MOLLOWE: Kit? Nay, it's Kitty now. What, is great Saxton so cowardly he fears the rage of Mollowe. [During the following speech, Mollowe backs Saxton into the cage.] Is great Saxton so passionate for being deprived of the joys of Mollowe's flesh? Learn thou of Kitty's manly fortitude, and scorn those joys thou never shalt possess. [Throwing the money at Saxton.] Go bear these triflings to your whoreson. Seeing Saxton hath already incurred eternal death by desperate acts against Kitty. I'll teach thee to behave thyself. [Kicking Saxton into the cage. Mollowe follows.] Letting you live in all voluptuousness, having thee ever to attend on me, to give whatsoever I shall ask, to do whatsoever I demand, and to always be obedient to my will. [Saxton rises, teary-eyed.] Too much of water hast thou?

SAXTON: Tears seven times salt burn out the sense and virtue of mine eye! I forbid my tears; but yet it is our trick; nature her custom holds, let shame say what it will.

MOLLOWE: Poor Saxton. Flowering tears thou shed. [Removing Saxton's clothes.] When these are gone, the woman will be out that pierced the fearful hollow of thine eye. My life, for you was just one night in jail.

SAXTON: Believe me, love, it was not just one night in jail. But in my bosom, to be to thee that night since a constant torchbearer to our love.

MOLLOWE: It so, I have more care to stay and let our deaths be one.

SAXTON: [Removing Mollowe's clothes.] Art thou back so? Love, lord, poet, ay husband, friend, spy, I must love thee in every hour.

The cage begins to ascend as they embrace

MOLLOWE: I again behold my rat. Ay, so conjured.

SAXTON: Ay, a cutpurse of the empire and the rule.

Notes

1. On transversal poetics, beyond this book, see Bryan Reynolds, "The Devil's House, 'or worse' " (*Theatre Journal* 49.2 [1997]: 143–67), *Becoming Criminal: Transversal Performance and Cultural Dissidence in Early Modern England* (Baltimore: Johns Hopkins University Press, 2002) and *Performing Transversally: Reimagining Shakespeare and the Critical Future* (New York: Palgrave Macmillan, 2003).

2. See Louis Althusser, *Lenin and Philosophy and Other Essays*, trans. Ben Brewster (New York: Monthly Review Press, 1971), 127–88.

3. From a classical Marxist standpoint of theorists like Nicos Poulantzas, Althusser's division of the state is problematic not only because of its ahistorical qualities, but also because it "diminishes the specificity of the *economic state apparatus* by dissolving it into the the various repressive and ideological apparatuses; it thus prevents us from locating the state network in which the power of the hegemonic fraction of the bourgeoisie is essentially centered" (*State, Power, and Socialism* 33).

4. Reynolds' conception of "state power" as any force from any source that works to consolidate social entities (toward the creation of a society) differs from that of classical Marxists like Nicos Poulantzas: "When we speak for example of *state power*, we cannot mean by it the mode of the state's articulation and intervention at the other levels of the structure; *we can only mean the power a determinate class* to whose interests (rather than to those of other classes) the state corresponds" (*Political Power and Social Classes*, 100).

5. While Laclau and Mouffe see self-identity as "post-Marxist" (4), as in their case, the term is typically applied to any anti-essentialist and/or anti-determinist reformulation of Marxism.

6. In their preface to the second edition of *Hegemony and Socialist Strategy* (London and New York: Verso, 2001), Laclau and Mouffe maintain that, "Given the magnitude of these epochal changes ["the end of the Cold War and the disintegration of the Soviet system"], we were surprised, in going through the pages of this not-so-recent book again, at how little we have to put into the question the intellectual and political perspective developed therein. Most of what has happened since then has closely followed the pattern suggested in our book, and those issues which were central to our concerns at that moment have become ever more prominent in contemporary discussions" (vii). Since, in their estimation, there is no need for them to update their theories so that they may better account for and engage with either sociopolitical changes or developments in scholarship since the mid 1980s, they conclude their preface with the declaration: "So our motto is: 'Back to the hegemonic struggle' " (xix).

7. See Gabrielle Schwab, "The Subject of the Political Unconscious," in Mark Poster ed., *Politics, Theory, and Contemporary Culture* (New York: Columbia University Press, 1993): 131–58; Kathleen Martindale, "Fredric Jameson's Critique of Ethical Criticism: a Deconstructed Marxist Feminist Response," *Feminist Critical Negotiations*, eds. Alice Parker and Elizabeth Meese (Amsterdam/Philadephia: John Benjamins Publishing, 1992), 33–43; and Pamela McCallum, "Question of Ethics: Reading Kathleen Martindale Reading Fredric Jameson" (*Resources for Feminist Research* 25.3–4 [1997]: 64–9).

8. See Anna Marie Smith, *Laclau and Mouffe: the Radical Democratic Imaginary* (London and New York: Routledge, 1998); Carol Stabile, "Feminism and the Ends of Postmodernism," *Materialist Feminism: a Reader in Class, Difference, and Women's Lives*, eds. Rosemary Hennessy and Chyrs Ingraham (New York and London: Routledge 1997), 395–458; Teresa Ebert, *Ludic Feminism and After: Postmodernism, Desire, and Labor in Late Capitalism* (Ann Arbor: University of Michigan Press, 1996); and Ellen Meiksins Wood, *The Retreat From Class: a New "True" Socialism* (London and New York: Verso, 1986).

9. The 1612 entry in *The Consistory of London Correction Book* cited here appears as Appendix E in Paul Mulholland's edition of Middleton and Dekker's *The Roaring Girl* (Manchester: Manchester University Press, 1987), 262–3. The entry is also reprinted in the introduction to *Counterfeit Ladies: the Life and Death of Mal Cutpurse and the Case of Mary Carleton*, eds. Janet Todd and Elizabeth Spearing (London: William Pickering, 1994), vii–liii; xiv–xv.

10. For Lacan, *points de caption* operate within a system of discourse, and a certain number of these quilting points are "necessary for a human being to be called normal, and which, when they are not established, or when they give way, make a psychotic" (268–9).
11. See Laclau and Mouffe (112). For Laclau and Mouffe, the "nodal point," which they also refer to as a "master-signifier" (xi), can also function like Derrida's conception of the "transcendental signified": " 'Man' is a fundamental nodal point from which it has been possible to proceed, since the eighteenth century, to the 'humanization' of a number of social practices" (117). For Derrida's similar, but more complexly theorized, concept of "supplementarity," see the chapter entitled "Structure, Sign, and Play in the Discourse of the Human Sciences" in his *Writing and Difference*, trans. Alan Bass (Chicago: University of Chicago Press, 1978), 278–94.
12. Among the more recent critical works on criminal culture in early modern England are Reynolds' *Becoming Criminal* and Linda Woodbridge's *Vagrancy, Homelessness, and English Renaissance Literature* (Urbana: University of Illinois Press, 2001).
13. "Shakespace" was first introduced in Donald Hedrick and Bryan Reynolds' "Shakespace and Transversal Power" in *Shakespeare Without Class: Misappropriations of Cultural Capital*, eds. Hedrick and Reynolds (New York: Palgrave, – now Palgrave Macmillan, 2000), 3–47.
14. While the conductors of state power function to create a hegemonic culture and "image of the totalized state," as Reynolds specifies, the "use of the term 'state machinery' should make explicit the multifarious and discursive nature of state power, and thus prevent the misperception of this dynamic as resultant from a conspiracy led by a monolithic state" ("Devil's House" 145).
15. For a detailed account of Reynolds' perspective on desire as both the subject and object of desire, see his "Becoming a Body Without Organs: the Masochistic Quest of Jean-Jacques Rousseau," *Deleuze and Guattari: New Mappings in Politics, Philosophy, and Culture*, eds. Kevin Jon Heller and Eleanor Kaufman (Minneapolis: University of Minnesota Press, 1998), 191–208.
16. Making an argument for Laclau and Mouffe's essentialism, Jacob Torfing sums up the apparently self-defeating ("double-edged sword") logic to their conceptualization of antagonism: "Hence, if antagonism is constitutive of all social identity, if there is always a constitutive outside that is both the condition of possibility and the condition of impossibility of any identity, there is an essential accidentalness that is constitutive of identity. However, if this accidentalness threatens an identity, that identity will be experienced as incomplete, as the vain aspiration to a fullness that will always escape it" (51–3).
17. For related transversal readings of *Macbeth*, *Othello*, and *The Tempest*, see Bryan Reynolds' "Untimely Ripped," Joseph Fitzpatrick, Bryan Reynolds, and Janna Segal's "Venetian Ideology or Transversal Power?: Iago's Motives and the Means by which Iago's Falls," and Bryan Reynolds and Ayanna Thompson's "Inspriteful Ariels: Transversal Tempests," all of which appear in Reynolds' *Performing Transversally*.
18. Bryan Reynolds and James Intriligator introduced the investigative-expansive mode of analysis in a paper entitled "Transversal Power," given at the

Manifesto Conference at Harvard University on May 9, 1998. For more on the "i.e. mode," see zooz, "Transversal Poetics: I.E. Mode" in *Performing Transversally*.

19. For more on "subjunctive space," see Reynolds' "Transversal Performance: Shakespace, the September 11th Attacks, and the Critical Future" in *Performing Transversally*: 1–29.

20. See Levin, "The Poetics and Politics of Bardicide" (*PMLA* 105.3 [1990]: 491–504), and Vickers, *Appropriating Shakespeare: Contemporary Critical Quarrels* (New Haven: Yale University Press, 1993).

21. For a chronology of Mary Frith's (1585–1659) criminal career, see Gustav Ungerer, "Mary Frith, Alias Moll Cutpurse, in Life and Literature" (*Shakespeare Studies* 28 [2000]: 42–84). According to Ungerer, Mary Frith (1585–1659) was first indicted on purse-snatching charges, along with two other women (Jane Hill and Jane Styles), in 1600 (62), and her "deviant behavior as a transvestite dates from about 1608" (55).

22. The January 27, 1612 entry in *The Consistory of London Correction Book* has been used by many critics and historiographers to establish that Mary Frith at least once appeared on stage, as advertised in the Epilogue to *The Roaring Girl*, at the Fortune Theatre following the performance of the play. In an oft-cited letter to Sir Dudley Carleton dated February 12, 1612, John Chamberlain recounts Mary Frith's drunken, disorderly public penance at St. Paul's Cross the previous Sunday (*The Letters of John Chamberlain* Vol. 1, ed. Norman Egbert McClure [Philadelphia: American Philosophical Society, 1939], 334). In addition to her theatrical representation as a character in *The Roaring Girl*, Moll Cutpurse was also a character who briefly appeared in Nathaniel Field's *Amends for Ladies* (1618). Moll Cutpurse's notoriety is also evident by the number of literary works published during her lifetime that made reference to her, including Thomas Dekker's *If It Be Not Good, the Devil Is In It* (1611/12), William Rowley, Thomas Dekker and John Ford's *The Witch of Edmonton* (1621), and Richard Brome's *The Court Beggar* (1640). As Mulholland notes, the now lost *Madde Pranckes of Mery Mall of the Banckside, with her Walks in Mans apparel, and to what Purpose* by John Day, entered on August 7, 1610 in the *Stationer's Register*, may be the earliest literary representation of Mary Frith (Introduction 13). "Biographical" accounts of Mary/Moll include *The Life and Times of Mrs. Mary Frith, Commonly called Mal Cutpurse* (1662) and Alexander Smith's *A Complete History of the Lives and Robberies of the Most Notorious Highwaymen, Shoplifts, and Cheats of Both Sexes* (1719). For a record of the various documentary sources, theatrical representations, and literary references to Mary Frith/Moll Cutpurse dating from the seventeenth to the eighteenth centuries, see Paul A. Mulholland's introduction to *The Roaring Girl: Thomas Middleton and Thomas Dekker*, ed. Mulholland (Manchester: Manchester University Press, 1987), 1–65; Randall S. Nakayama's introduction to *The Life and Death of Mrs. Mary Frith, Commonly Called Moll Cutpurse*, ed. Nakayama (New York: Garland Press, 1993), vii–xxix; Janet Todd and Elizabeth Spearing's introduction to *Counterfeit Ladies: The Life and Death of Mal Cutpurse, The Case of Mary Carleton*; and Gustav Ungerer's "Mary Frith, Alias Moll Cutpurse, in Life and Literature."

23. For a debate over the authenticity of *The Life and Death of Mrs. Mary Frith, Commonly Called Mal Cutpurse*, see Ungerer's "Mary Frith, Alias Moll Cutpurse." Critical of Todd and Spearing's assessment of the biography as "the only [text] that gives anything like an account of the actual woman rather than a mythical figure, or that could derive from information given by the original 'Moll,' Mary Frith herself" (Todd and Spearing x), Ungerer argues that the "fragmentary" biography is "male-orientated," and that the work offers only a fictionalized account intended for public consumption: "[t]he biographers were committed to adjusting their subject in conformance to the stereotypical criminal of fictional biography" (42).

24. T. S. Eliot's "Thomas Middleton" first appeared in the *Times Literary Supplement*, June 30, 1927. The citation is here taken from Eliot's *Elizabethan Dramatists* (New York: Haskell House, 1964), 87–100; 99.

25. For a survey of seven revivals of *The Roaring Girl* produced between 1951 and 1983, see Paul A. Mulholland's "Let Her Roar Again: *The Roaring Girl* Revived" (*Research Opportunities in Renaissance Drama* 28 [1985]: 15–27), and Mullholland's introduction to his edition of *The Roaring Girl*. Among the seven productions Mulholland discusses, two date from the 1970s, including Sue-Ellen Case's 1979 production at the University of California at Berkeley, and three from the early 1980s, among which is the 1983 RSC production starring Helen Mirren as Moll. It is Mulholland's contention that while the most recent productions he considers have used the source text as a platform to address feminist issues, the play itself does not support a feminist reading, and must be altered (and adulterated, in his opinion) to produce a more contemporary message concerning gender politics ("Let Her Roar Again" 25; Introduction 53).

26. Articles on *The Roaring Girl* appearing shortly before the publication of Rose's piece include two works by Paul A. Mulholland, "The Date of *The Roaring Girl*" (*The Review of English Studies* 28 [1977]: 18–31) and "Some Textual Notes on *The Roaring Girl*" (*The Library* 32 [1977]: 333–43), both of which focus on intertextual evidence that can be used to historicize the play. Other pre-Rose works that offer a more critical analysis of the play include Patrick Cheney's "Moll Cutpurse as Hermaphrodite in Dekker and Middleton's *The Roaring Girl*" (*Renaissance and Reformation* 7.2 [1983]: 120–34), in which Cheney identifies Moll's "paradoxical" portrayal as stemming from her roots in the hermaphroditic Platonic and Neoplatonic tradition popularized by Spenser. Comparing *The Roaring Girl* to Spenser and others, Cheney argues that Moll is a hermaphroditic ideal synthesis of male and female traits, a unity of "opposites" that, functioning as a "symbol" for combinations of genre, authorial, and sexual identity formations, has transgressive social, personal, and artistic potential (132). His reading of Moll as a unity of "opposites" whose opposing forces negate sexual desire (130) could be read as an affirmation of a binary gender system that disallows desire to those who cross its constructed paths. Nonetheless, while Cheney's article is generally unacknowledged by post-Rose critics re-reading *The Roaring Girl*, his piece appears to be the first published scholarly article to consider sexuality and gender formation in Middleton and Dekker's play. For a brief consideration of *The Roaring Girl* published prior to the

publication of Rose's piece, also see Linda Woodbridge's *Women and the English Renaissance: Literature and the Nature of Womankind, 1540–1620* (Urbana: University of Illinois Press, 1984), in which Woodbridge labels the playwrights' recreation of the "real" as a "favourable treatment to a man-clothed virago" (250).

27. In 2001, the Shenandoah Shakespeare Company produced *The Roaring Girl*, contextualizing it topically with current politics and popular trends. For instance, the canting was done like a rap song. In the words of Craig Dionne, who saw the production, "The point was to show Moll as a transhistorical object of vexed desire via pop culture" (personal email correspondence, 12/8/2002).

28. Of the thirty recently published works on Mary Frith/Moll Cutpurse considered here, only three focus primarily on a Mary Frith of biographical and/or historical record rather than on the Moll Cutpurse of theatrical fare: Nakayama's introduction to *The Life and Death of Mrs. Mary Frith*; Todd and Spearing's introduction to *Counterfeit Ladies*; and Ungerer's "Mary Frith, Alias Moll Cutpurse." While some recent scholarship, such as Ungerer's, has sought to retrieve Mary from the Moll of the theatrical stage and/or the Moll of the biographical page by focusing on the Mary Frith of historiographical "fact," most critics, rather than contending with archival-based conjectures, focus on Middleton and Dekker's projection of the public persona, referencing the 1612 record of Mary Frith's arrest for her onstage appearance at the Fortune Theatre and/or Chamberlain's 1612 letter recounting Frith's public penance to legitimate the transgressive potential of the "real" Moll Cutpurse. Jane Baston, for instance, contrasts archival accounts of Frith with her theatricalicalized counterpart to construct her argument that while Mary Frith was a transgressive figure who subverted the patriarchal order through her cross-dressing and public penance, Middleton and Dekker's "sanitized Moll" is orchestrated to "subtly undercut her political potency" ("Rehabilitating Moll's Subversion in *The Roaring Girl*," *Studies in English Literature 1500–1900* 37.2 [1997]: 317–35; 323). Jean Howard, on the other hand, references the 1612 entry documenting Frith's arrest for lute-playing cross-dressed at the Fortune Theatre to argue that the theatrical Moll is more transgressive than the real. Howard argues that while "[t]he original Moll [...] was transgressive for playing her lute on the public stage," Middleton and Dekker's Moll "is even more transgressive in that her instrument is not the lute, able to be tucked decorously beneath the breast, but the viol, played with legs akimbo. Moreover, she seems to appropriate this instrument not so much to make herself an erotic object, as to express her own erotic subjectivity" ("Sex and Social Conflict: the Erotics of *The Roaring Girl*," *Erotic Politics: Desire on the Renaissance Stage*, ed. Susan Zimmerman [New York: Routledge, 1992], 170–90; 184).

29. For considerations of *The Roaring Girl* as an unconventional text in its breaching of genre and authorship conventions, see, among other works, Cheney, "Moll Cutpurse as Hermaphrodite in Dekker and Middleton's *The Roaring Girl*"; Mary Beth Rose, "Women in Men's Clothing: Apparel and Social Stability in *The Roaring Girl*" (*English Literary Renaissance* 14.3 [1984]: 367–91); Viviana Comensoli, "Play-making, Domestic Conduct, and the Multiple Plot in *The Roaring Girl*" (*Studies in English Literature 1500–1900*

27.2 [1987]: 249–66); and Valerie Forman, "Marked Angels: Counterfeits, Commodities, and *The Roaring Girl*" (*Renaissance Quarterly* 54.4 [2002]: 1531–60).

30. For feminist/queer readings of Middleton and Dekker's version of Moll Cutpurse as a progressive transgressor of patriarchal gender, class, and sexual structures, see, among others, Rose, "Women in Men's Clothing: Apparel and Social Stability in *The Roaring Girl*"; Comensoli, "Play-making, Domestic Conduct, and the Multiple Plot in *The Roaring Girl*"; Jo E. Miller, "Women and the Market in *The Roaring Girl*" (*Renaissance and Reformation* 19.1 [1990]: 11–23); Jonathon Dollimore, *Sexual Dissidence: Augustine to Wilde, Freud to Foucault* (Oxford: Clarendon Press, 1991), 293–9; Howard, "Sex and Social Conflict: the Erotics of *The Roaring Girl*"; Stephen Orgel, "The Subtexts of *The Roaring Girl*" (*Erotic Politics: Desire on the Renaissance Stage*, ed. Susan Zimmerman [New York: Routledge, 1992]), 12–26; Jodi Mikalachki, "Gender, Cant, and Cross-Talking in *The Roaring Girl*" (*Renaissance Drama* 25 [1994]: 119–43); Lloyd Edward Kermode, "Destination Doomsday: Desire for Change and Changeable Desires in *The Roaring Girl*" (*English Literary Renaissance* 27.3 [1997]: 421–42); Craig Rustici, "The Smoking Girl: Tobacco and the Representation of Mary Frith" (*Studies in Philology* 96.2 [1999]: 159–79); Herbert Jack Heller, *Penitent Brothellers: Grace, Sexuality, and the Genre in Thomas Middleton's City Comedies* (Newark: University of Delaware Press, 2000), 151–70; and Reynolds, *Becoming Criminal*, 64–94. Feminist/queer/materialist/new historicist critics who read the theatrical Moll as an affirmation of the dominant order and position the "real" Moll Cutpurse as the "truly" dissident figure include, among others, Susan E. Krantz, "The Sexual Identity of Moll Cutpurse in Dekker and Middleton's *The Roaring Girl* and in London" (*Renaissance and Reformation* 19.1 [1995]: 5–20); and Baston, "Rehabilitating Moll's Subversion in *The Roaring Girl*." For feminist/queer/materialist and/or (new) historicist readings of Middleton and Dekker's Moll as a critically misread figure, see, among others, Mulholland's "Let Her Roar Again: *The Roaring Girl* Revived," and Mullholland's introduction to his edition of *The Roaring Girl*; Marjorie Garber, "The Logic of the Transvestite: *The Roaring Girl* (1608)" (*Staging the Renaissance: Reinterpretations of Elizabethan and Jacobean Drama*, eds. David Scott Kasdan and Peter Stallybrass [New York: Routledge, 1991]), 221–34; Deborah Jacobs, "Critical Imperialism and Renaissance Drama: the Case of *The Roaring Girl*" (*Feminism, Bakhtin, and the Dialogic*, eds. Dale M. Bauer and Susan Jaret McKinstry [Albany: State University of New York Press, 1991]), 73–84; Krantz, "The Sexual Identity of Moll Cutpurse in Dekker and Middleton's *The Roaring Girl* and in London"; Baston, "Rehabilitating Moll's Subversion in *The Roaring Girl*"; and Forman, "Marked Angels: Counterfeits, Commodities, and *The Roaring Girl*." For a reading of Dekker and Middleton's Moll as a figure affirming a hierarchical class structure, see William West, "How to Talk the Talk, or, The Work of Cant on the Jacobean Stage" (*English Literary Renaissance* 33 (2003): 228–51. For historiographical accounts of Mary Frith that, focusing on the 1662 biography, cast the "real" Moll Cutpurse as a dissident figure, see Naka-yama's introduction to *The Life and Death of Mrs. Mary Frith, Commonly Called Moll Cutpurse*; and Todd and Spearing's introduction to *Counterfeit*

Ladies: the Life and Death of Mal Cutpurse, The Case of Mary Carleton. For a historiographical critique of mystifications of the "real" Mary Frith based on assumptive readings of the 1662 biography as "fact," see Ungerer, "Mary Frith, Alias Moll Cutpurse, in Life and Literature."

31. For a critique of postmodern investment in the English Renaissance as "Early Modern," see Margreta de Grazia, Maureen Quilligan, and Peter Stallybrass' introduction to *Subject and Object in Renaissance Culture*, eds. de Grazia, Quilligan, and Stallybrass (Cambridge: Cambridge University Press, 1996), 1–13; and de Grazia's "The Ideology of Superfluous Things: *King Lear* as Period Piece," *Subject and Object in Renaissance Culture*, 17–42.

32. For a reading of Faustus' malcontention and transgressive power, see Jonathon Dollimore, *Radical Tragedy: Religion, Ideology and Power in the Drama of Shakespeare and his Contemporaries* (Sussex: Harvester Press, 1984).

33. Those critics that cite 3.1 from *The Roaring Girl* in support of their arguments include Cheney (1983); Rose (1984); Comensoli (1987); Miller (1990); Dollimore (1991); Garber (1991); Jacobs (1991); Howard (1992); Orgel (1992); Mikalachki (1994); Krantz (1995); Baston (1997); Kermode (1997); Rustici (1999); and Heller (2000).

Works cited

Althusser, Louis. *Lenin and Philosophy and Other Essays.* Trans. Ben Brewster. New York: Monthly Review Press, 1971.

Baston, Jane. "Rehabilitating Moll's Subversion in *The Roaring Girl." Studies in English Literature 1500–1900* 37.2 (1997): 317–35.

Butler, Judith. *Gender Trouble: Feminism and the Subversion of Identity.* London: Routledge, 1990.

Cheney, Patrick. "Moll Cutpurse as Hermaphrodite in Dekker and Middleton's *The Roaring Girl." Renaissance and Reformation* 7.2 (1983): 120–34.

Comensoli, Viviana. "Play-making, Domestic Conduct, and the Multiple Plot in *The Roaring Girl." Studies in English Literature 1500–1900* 27.2 (1987): 249–66.

de Grazia, Margreta. "The Ideology of Superfluous Things: *King Lear* as Period Piece." *Subject and Object in Renaissance Culture.* Eds. Margreta de Grazia, Maureen Quilligan, and Peter Stallybrass. Cambridge: Cambridge University Press, 1996. 17–42.

——— Maureen Quilligan, and Peter Stallybrass. Introduction. *Subject and Object in Renaissance Culture.* Eds. Margreta de Grazia, Maureen Quilligan, and Peter Stallybrass. Cambridge: Cambridge University Press, 1996. 1–13.

Dekker, Thomas and Thomas Middleton. *The Roaring Girl. English Renaissance Drama: a Norton Anthology.* Eds. Katherine Eisaman Maus and David Bevington. New York: Norton, 2002. 1377–451.

Derrida, Jacques. *Writing and Difference.* Trans. Alan Bass. Chicago: University of Chicago Press, 1978.

Dionne, Craig. "Fashioning Outlaws: the Early Modern Rogue and Urban Culture." *Rogues and Early Modern English Culture.* Eds. Craig Dionne and Steve Mentz. Ann Arbor: University of Michigan Press, 2004. 33–61.

Dollimore, Jonathan. *Radical Tragedy: Religion, Ideology and Power in the Drama of Shakespeare and his Contemporaries.* Sussex: Harvester Press, 1984.
—— *Sexual Dissidence: Augustine to Wilde, Freud to Foucault.* Oxford: Clarendon Press, 1991.
Eliot, T. S. "Thomas Middleton." *Elizabethan Dramatists.* New York: Haskell House, 1964. 87–100.
Fitzpatrick, Joseph, and Bryan Reynolds. "Venetian Ideology or Transversal Power?: Iago's Motives and the Means by which Iago Falls." *Othello: New Critical Essays.* Ed. Philip C. Kolin. New York: Routledge, 2002. 203–19.
—— "Venetian Ideology or Transversal Power?: Iago's Motives and the Means by which Iago Falls." *Performing Transversally: Reimagining Shakespeare and the Critical Future.* Bryan Reynolds. New York: Palgrave Macmillan, 2003.
Forman, Valerie. "Marked Angels: Counterfeits, Commodities, and *The Roaring Girl.*" *Renaissance Quarterly* 54.4 (2002): 1531–60.
Foucault, Michel. *The History of Sexuality: an Introduction,* Vol. 1. New York: Random House, 1990.
Galford, Ellen. *Moll Cutpurse: Her True History.* Ithaca: Firebrand, 1985.
Garber, Marjorie. "The Logic of the Transvestite: *The Roaring Girl* (1608)." *Staging the Renaissance: Reinterpretations of Elizabethan and Jacobean Drama.* Eds. David Scott Kasdan and Peter Stallybrass. New York: Routledge, 1991. 221–34.
—— *Vested Interests: Cross-Dressing & Cultural Anxiety.* New York & London: Routledge, 1992.
Hedrick, Donald, and Bryan Reynolds. "Shakespace and Transversal Power." *Shakespeare Without Class: Misappropriations of Cultural Capital.* Eds. Donald Hedrick and Bryan Reynolds. New York: Palgrave Macmillan, 2000. 3–47.
Heller, Herbert Jack. *Penitent Brothellers: Grace, Sexuality, and the Genre in Thomas Middleton's City Comedies.* Newark: University of Delaware Press, 2000.
Hopkins, D. J., and Bryan Reynolds. "The Making of Authorships: Transversal Navigation in the Wake of *Hamlet,* Robert Wilson, Wolfgang Wiens, and Shakespace." *Shakespeare After Mass Media.* Ed. Richard Burt. New York: St. Martin's Press, 2002. 265–86.
—— "The Making of Authorships: Transversal Navigation in the Wake of *Hamlet,* Robert Wilson, Wolfgang Wiens, and Shakespace." *Performing Transversally: Reimagining Shakespeare and the Critical Future.* New York: Palgrave Macmillan, 2003.
Howard, Jean E. "Sex and Social Conflict: the Erotics of *The Roaring Girl.*" *Erotic Politics: Desire on the Renaissance Stage.* Ed. Susan Zimmerman. New York: Routledge, 1992. 170–90.
Jacobs, Deborah. "Critical Imperialism and Renaissance Drama: the Case of *The Roaring Girl.*" *Feminism, Bakhtin, and the Dialogic.* Eds. Dale M. Bauer and Susan Jaret McKinstry. Albany: State University of New York Press, 1991. 73–84
Jameson, Fredric. *The Political Unconscious: Narrative as a Socially Symbolic Act.* Ithaca: Cornell University Press, 1981.
Kermode, Lloyd Edward. "Destination Doomsday: Desire for Change and Changeable Desires in *The Roaring Girl.*" *English Literary Renaissance* 27.3 (1997): 421–42.

Krantz, Susan E. "The Sexual Identity of Moll Cutpurse in Dekker and Middleton's *The Roaring Girl* and in London." *Renaissance and Reformation* 19.1 (1995): 5–20.

Lacan, Jacques. *The Psychoses*. Trans. Russell Grigg. New York: W. W. Norton, 1993.

Laclau, Ernesto. *New Reflections on the Revolution of Our Time*. London and New York: Verso, 1990.

Laclau, Ernesto, and Chantal Mouffe. *Hegemony and Socialistic Strategy: Towards a Radical Democratic Politics*. London and New York: Verso, 1985.

Levin, Richard. "The Poetics and Politics of Bardicide." *PMLA* 105.3 (1990): 491–504.

Maus, Katherine Eisaman. Introduction. "The Roaring Girl." *English Renaissance Drama: a Norton Anthology*. Eds. Katherine Eisaman Maus and David Bevington. New York: Norton, 2002. 1371–76.

Mikalachki, Jodi. "Gender, Cant, and Cross-Talking in *The Roaring Girl*." *Renaissance Drama* 25 (1994): 119–43.

Miller, Jo E. "Women and the Market in *The Roaring Girl*." *Renaissance and Reformation* 19.1 (1990): 11–23.

Mulholland, Paul A. "The Date of *The Roaring Girl*." *The Review of English Studies* 28 (1977): 18–31.

—— "Some Textual Notes on *The Roaring Girl*." *The Library* 32 (1977): 333–43.

—— "Let Her Roar Again: *The Roaring Girl* Revived." *Research Opportunities in Renaissance Drama* 28 (1985): 15–27.

—— Introduction. *The Roaring Girl: Thomas Middleton and Thomas Dekker*. Ed. Paul A. Mulholland. Manchester: Manchester University Press, 1987. 1–65.

Nakayama, Randall S. Introduction. *The Life and Death of Mrs. Mary Frith, Commonly Called Moll Cutpurse*. Ed. Randall S. Nakayama. New York: Garland Press, 1993. vii–xxix.

Orgel, Stephen. "The Subtexts of *The Roaring Girl*." *Erotic Politics: Desire on the Renaissance Stage*. Ed. Susan Zimmerman. New York: Routledge, 1992. 12–26.

Poulantzas, Nicos. *Political Power and Social Classes*. London: NLB and Sheed & Ward, 1973.

—— *State, Power, Socialism*. London and New York: Verso, 2000.

Reynolds, Bryan. "The Devil's House, 'or worse': Transversal Power and Antitheatrical Discourse in Early Modern England." *Theatre Journal* 49.2 (1997): 143–67.

—— "Becoming a Body without Organs: the Masochistic Quest of Jean-Jacques Rousseau." *Deleuze and Guattari: New Mappings in Politics, Philosophy, and Culture*. Eds. Eleanor Kaufman and Kevin Jon Heller. Minneapolis: University of Minnesota Press, 1998. 191–208.

—— *Becoming Criminal: Transversal Performance and Cultural Dissidence in Early Modern England*. Baltimore: Johns Hopkins University Press, 2002.

—— *Performing Transversally: Reimagining Shakespeare and the Critical Future*. New York: Palgrave Macmillan, 2003.

Rose, Mary Beth. "Women in Men's Clothing: Apparel and Social Stability in *The Roaring Girl*." *English Literary Renaissance* 14.3 (1984): 367–91.

Rustici, Craig. "The Smoking Girl: Tobacco and the Representation of Mary Frith." *Studies in Philology* 96.2 (1999): 159–79.

Smith, Paul. *Discerning the Subject*. Minneapolis: University of Minnesota Press, 1988.

Todd, Janet, and Elizabeth Spearing. Introduction. *Counterfeit Ladies: the Life and Death of Mal Cutpurse, The Case of Mary Carleton*. Eds. Janet Todd and Elizabeth Spearing. London: William Pickering, 1994. vii–liii.

Ungerer, Gustav. "Mary Frith, Alias Moll Cutpurse, in Life and Literature." *Shakespeare Studies* 28 (2000): 42–84.

Vickers, Brian. *Appropriating Shakespeare: Contemporary Critical Quarrels*. New Haven: Yale University Press, 1993.

West, William, "How to Talk the Talk, or, The Work of Cant on the Jacobean Stage," *English Literary Renaissance* 33 (2003): 228–51.

Woodbridge, Linda. *Women and the English Renaissance: Literature and the Nature of Womankind, 1540–1620*. Urbana: University of Illinois Press, 1984.

——— *Vagrancy, Homelessness, and English Renaissance Literature*. Urbana: University of Illinois Press, 2001.

Žižek, Slavoj. *The Sublime Object of Ideology*. London and New York: Verso, 1989.

zooz. "Transversal Poetics: I. E. Mode." *GESTOS: Teoría y Práctica del Teatro Hispánico* 18:35 (April 2003).

3
The Delusion of Critique: Subjunctive Space, Transversality, and the Conceit of Deceit in *Hamlet*

Anthony Kubiak & Bryan Reynolds

In their recent essay, " 'A little touch of Harry in the night': Translucency and Projective Transversality in the Sexual and National Politics of *Henry V*," Donald Hedrick and Bryan Reynolds argue that Shakespeare's Princess Catherine potentially undermines King Henry's fantasized domination of her during sex by occupying antithetically the conceptual-emotional spacetime of Catherine's blindness by "winking" (5.2.262). In other words, Henry's fantasy of Catherine closing her eyes during sex that he shares with Burgundy ("Yet they do wink and yield, as love is blind and enforces" [5.2.259]) so that he can enter her from behind ("and so I shall catch the fly, your cousin [Catherine], in the latter end, and she must be blind too" [5.2.270–1]), inadvertently makes room for Catherine, in her imagination, to "disappear" and thereby deceive Henry. As Hedrick and Reynolds put it,

> Sex-without-seeing, from the perspective of this [Henry's] fantasy, indicates a trajectory of male and national domination. But a transversal reading suggests a different possibility altogether – a key to the scene, if not to the entire play: by closing one's eyes, one "disappears" the other, or even transforms the other into someone else. Instead of transversality as a becoming-other of one's own subjectivity, becoming what you are not, transversality might be now thought of in terms of transforming the other outside himself or herself, a projective transversality or "Renaissance other-fashioning." What we are suggesting is that transversality and translucency may act as mechanisms with an entirely different outcome or purpose for Catherine than for Henry in his performance of himself. (*Performing Transversally* 175)

"Disappearing" in *Hamlet*, however, is an enterprise of an entirely different character. Unlike in *Henry V*, where winking reveals irony and empowers, in *Hamlet* a wink in the face of the play's deceptions can be fatal. One cannot "disappear" what one cannot know, except for those doing the plotting successfully in what Anthony Kubiak has termed the "Hamletic guise." In fact, in the play, the retributive violence that participates in various "disappearings" is also presented in Hamletic guise; that is to say, vengeance is *plotted* like the play itself (or any play for that matter) through the machinations of fabrication and deceit, such as feigned madness, the secret taint of the swords, the hidden poisoning of drink. By *Hamlet*'s final act, deceit and vengeance meld into the same; deceit becomes the cause of violence, and violence the agency of deceit.

Behind all deceit, particularly in *Hamlet*, according to the Hamletic model we are proffering here, is the envenomed, plotting character of thought, played out in what transversal theory refers to as "subjunctive space," the hypothetical space of both "as ifs" and "what ifs" operative in between "subjective territory" (the conceptual-emotional-physical range from which a given subject perceives and experiences) and "transversal territory" (the nonsubjectified, transforming space through which people journey when they defy or surpass the boundaries of their subjective territory).[1] The profound relationship between fabrication and "subjunctive movement" implies a necessity in the organic patterns of thinking that gives rise to emplottedness, deceit, and violence – a kind of murderous Kantian *a priori*. It is as if in some overarching trajectory of natural history, beyond the pale of mere species evolution or accountable spacetime, the primal/primate mind has framed and indeed authored itself as tragedy. To contextualize this idea, we follow Reynolds's understanding of tragedy:

> Several factors determine the extent to which an action or event [such as thinking or a thought] is a tragedy. First of all, as with any performance (defined here as a self-consciously presented expression for an intended audience),[2] how the event is framed needs be taken into consideration. Framing refers to the aesthetic, social, cultural, political, ideological, and historical context established for the performance, whether it takes place in a stage-play or during a sports game or at a particular venue, like in a courtroom, university classroom, chapel, theater, or a pub. The sociopolitical conductors that work to instill our biases and predilections shape our relationship to that context and thus to what is expected to happen there. In most

cases, when the framing is less apparent and less understood, the potential for a tragedy-producing performance is greater. In other words, the more sudden and unexpected the injuring action, the more tragic potential it has. But framing also has to do with invest-ment. The degree to which the audience invests itself in the damag-ing event and its victims emotionally, conceptually, physically, financially – in terms of spacetime, energy, and emotion – directly affects the potential for tragedy; the greater the investment, the greater the potential for loss, the greater the potential for tragedy. (*Performing Transversally* 13)

Hence, framed within this rather tragic view of humanity's investiture in itself as a high-risk endeavor combined with the idea of a "deceitful imperative" is what could be construed as a "natural" emergence, a conception of tragedy that moves outside of the realm of literary form, convention, or genre, and comes to represent the structure of thought "comings-to-be" through the biosocial necessity of deceit. This comings-to-be of thought is, moreover, "framed" in/by mind – the thinking that "makes it so." Yet this "frame of mind" must always remain unseen, unknowable. The "frame of mind" that turns some-thing into theater is precisely the "language game" (following Wittgen-stein), or the Unconscious (following Freud) that can never be seen in mind – the unseen and unseeable proscenium.

In this formulation, deceit best precipitates success when least expected, especially if misleading or injuring people – physically and psychically – is, in part, the measure of efficacy (we must, of course, remember that the other measure of its efficacy is how much harmony and stability is retained in the social unit through the uses of deceit). When the subject, in this case human thought, loses control during processes of "becomings-other" and becomes more of/or something else than anticipated and/or desired, comings-to-be are occurring. According to transversal theory, becomings-other are active engage-ments, usually self-inaugurated and pursued intentionally, whereas comings-to-be, however spurred by becomings, are generated by the energies, ideas, people, societies, and so on to which the subject aspires, is lured toward, or encounters accidentally. Empathy and assimilation are no longer self-consciously accomplished. Instead, they happen unwittingly and unconsciously through ideological, emo-tional, and/or physiological adaptation and change. Transversal theory refers to the dynamics that cause and are caused by the linked pro-cesses of comings-to-be and becomings as "pressurized belongings."[3]

For a subject to become a member of an alternative group, to incorporate the qualities of a different subjective territory and/or to operate within a foreign official territory, assimilation and expulsion must occur at the expense of aspects of these destinations. The transformed subject causes overflow, expansion, or reconfiguration such that not all of the extant elements can remain or remain the same if the system is to maintain equilibrium.

The point here, however, relates to the selection by humans, however inadvertently, of the specific play, *Hamlet*, as the bellwether play of lies, subjunctivity, and pressurized belongings. Although over-critiqued and palimpsested throughout literary history, *Hamlet* remains framed as The Play in the Euro-American history of performance. It is the play of the world's most celebrated poet, Shakespeare, that is most invested in by dominant sociopolitical conductors (familial, educational, juridical, and governmental structures) of the state machinery (the amalgamation of the conductors that work to foster a cohesive society) of all Western societies because, we would argue, it deceives deceit itself. In effect, *Hamlet* creates a window through which people can become and come-to-be, among other things, Hamletic. *Hamlet* thus exemplifies what transversal theory calls "deceit conceits," which are clever schemes involving artifice and fiction performed in order to fracture, transform, and/or expand the conceptual and/or emotional range of an individual or individuals, which is to say, a targeted subjective territory or subjective territories, such as those occupied by many of *Hamlet*'s audiences throughout history (see Chapter 4). These are audiences who have engaged, to some degree, in the particular "articulatory space" – a discursive, multi-dimensional conceptual-emotional interface – that Donald Hedrick and Bryan Reynolds have termed "Shakespace,"[4] and the sometimes subset, sometimes superset, but nevertheless always related, "Hamletspace." Simply put, Shakespeare's play, *Hamlet*, is brilliantly, strategically transversal.

Else and or? Spaces of subjunctivity

Indeed, in Shakespeare's Elsinore, the setting for the action of *Hamlet*, the manipulations, deceits, ironies, and performances – the subjunctive and transversal movements – are multifarious. The character Hamlet pretends madness so that he might uncover the secret of his father's murder (recalling Henry's pretending in *1 Henry IV* to be a genuine member of Falstaff's gang only to become the secret of his own success). But the madness is no mere pretense, or rather, the pretense is itself pretended: "I am but mad north-north-west: when the wind is southerly I know a hawk from a handsaw" (2.2.374–5). His social performance, in other words, is an amalgam of remembering (to act mad) and forgetting (that he is acting), then remembering again (that he is forgetting), *ad infinitum*. Here remembering (to act), and forgetting (that one is acting) are both equally indispensable functions of consciousness, that is, regardless of the subject's subjective territory, certain performances must occur for the subject to be aware of anything, much less itself, operating anywhere in spacetime. Remembering and forgetting are not successes or failures of consciousness, but consciousness in action, performed, sustained.

This is why, in our view, theater is appropriate as a model for understanding consciousness; consciousness depends on theatricality, the continual framing, unframing, and reframing of performances, such as of remembering and of forgetting that one is acting in a particular spacetime; and theater, structurally, reflects consciousness, with its multiple framings and vistas of experience. Theater, then, the communal product of humans, and not the "Cartesian theater" often imagined as central to the organization of an individual's brain, is phenomena of interactive liveness, sentience, and awareness more apt to understanding consciousness than, say, computer models, such as Stephen Blaha's "classical probabilistic account,"[5] or Daniel Dennett's "multiple drafts model," neither of which consider phenomena of reactive and communicative reciprocity among the brain and entities beyond its own material limits.

Getting back to *Hamlet*, Claudius the killer, remembering and forgetting, reveals himself concealing his crime: "My words fly up, my thoughts remain below: / Words without thoughts never to heaven go" (3.3.98–9). Aware of, if nothing else, the inappropriateness of Hamlet's performance, Gertrude conceals her complicity (or seeming complicity, if any) by castigating Hamlet in his grief, "Why seems it so particular with thee?" (1.2.74). She launches her query as if death, indeed regicide and fratricide, had not come to Elsinore, and she thus enjoys still untainted the office of queen, or at least must appear so, having forgot the loss of her husband, and remembering that she has a new one, while all along she remembers to act the role of queen, and forgets any grief that might distract her from this performance. The Machiavellian tricksters Rosencrantz and Guildenstern, subjected to Hamlet's counter machinations, are themselves tricked to death, while Ophelia is humiliated and confounded by Hamlet's seeming and seamy accusations of infidelity; she is thereby pushed into an all-consuming transversality.[6]

Hamlet sets up the play within the play by which he will "catch the conscience of the king," and the play within *that* play, a dumb show, that doubly inscribes the murderous deed (presumably) done. Theater frames consciousness by which the theatrical world of the play is exposed, made cognizant. A fun house of mirrors that can be outlined, as Herbert Blau has shown, like this: Hamlet sets into motion – precipitating movements among the audience that are recollective, subjunctive, and/or transversal – by means of the play (both large and small) a theatrical panopticon of universal surveillance – Claudius watching the play, Gertrude watching Claudius watch the play, Hamlet watching Gertrude watching Claudius watch the play, Ophelia watching Hamlet, *ad infinitum, ad nauseum*: all speculations and meanings drowned in the concentric wash of the watching.[7] All the while, we, the audience, watch, and are watched, and so on (by the theater's security, our peers, other audience members, the state machinery ...).

Yet these are the more obvious manifestations or symptoms of deceit, the possibility of deceit (the performative deceit) being the immediate cause for the compulsion for surveillance, for the insecurity, uncertainty, and mistrust that deceit breeds. *Hamlet's* theater is woven of lies given over to our scrutiny, and lies concealed (as we will see later in one of Hamlet's more famous soliloquies). In seducing us with the obvious lie, Shakespeare slips the deeper deceit in the backdoor, so to speak. Thus, while we congratulate ourselves for our perspicacity, we are blindsided by Hamlet who lies to our faces. Finally, there are those lies that exist in a kind of liminal space: Hamlet feigns madness, but then sees ghosts no one else can see, or Hamlet leads Polonius to see whales and weasels where there are none, then leads Horatio to imagine the king/no-king illusively "frighted with false fire" (3.2.260). We come to suspect that deceit, employed here with purpose as deceit conceit, is more than a human instrumentality; it is an agent of action in the play (personified in Claudius, Hamlet, Rosencrantz, etc.), more than a social or political strategy: *it is an initial state of mind that needs resistance*. Yet whatever honesty does appear is merely honesty grafted onto lies, lies becoming the bedrock of perception (causing another king, Oedipus, to rake out his eyes), and truth nothing but a hallucinatory touchstone grounding the inevitability of deceit. Hence, the play's deceit conceits are, in salvo, a *tour de force* in projective transversality, which is to say that they powerfully promote transversal movements in all subjective territories exposed to the play, even those of the characters within it. Deceit, as the play's ethos, is, in other words, not a perversion of human conduct, but its first principle. We are born to deceit, *Hamlet* suggests, and we must learn the strategies of truth-telling, or rather, we must *unlearn* the Darwinian impulse to lie. Here, we would suggest, transversality would also appear in the extraordinary perversity of truth-telling, resisting the genetic predisposition for deceit.

To be sure, an effect of the many forms and degrees of deceit conceits perpetrated within and by *Hamlet* is to make us – as we are made to move transversally – suspect that there is something radically amiss "in this distracted globe" (1.5.97), in the The Globe itself, within and beyond the worlds of the play, and in that special space of phantoms that gives the stage its birth: the brain. Brain and stage become one not through the manifestation of the true, the beautiful, the tragic, the imaginary, the artistic, but through the very architectures of deceit, and perhaps no demonstration of deceit in the play is as calculating and nefarious as this:

> Now I am alone.
> O, what a rogue and peasant slave am I!
> Is it not monstrous that this player here
> But in a fiction, in a dream of passion,
> Could force his soul so to his own conceit
> That from her working all his visage wann'd,
> Tears in his eyes, distraction in's aspect,
> A broken voice, and his whole function suiting
> With forms to his conceit? and all for nothing!
> (2.2.543–51)

In this aside, the most direct invocation to a Coleridgean "willing suspension of disbelief," there seems to be a contract or trust between actor and audience that runs against the grain of the obvious: it operates in violation of "ocular proof" – so elusive in *Othello* (3.3.366). The actor is, after all, standing on stage, his fellow actors waiting in the wings, cueing on his words, in a theater filled with spectators, all eyes on him. And what is presumed to be an introspective recitative expression must be heard in the back rows: the aside, in other words, like the fourth wall or proscenium in modern theater, is a convention, a lie universally recognized and accepted, and so necessarily invisible and forgotten.

But in this case, the convention is complicated beyond its normal "confessional" mode. The Beckettian "all for nothing," an emotional Black Monday in the collapse of psychic economy, gives us pause, thinking back through the speech in memory: "Is it not monstrous that this player here ..." But wait. Which player where? Is Hamlet indicating the recently departed First Player (who is certainly not *here*, but rather *there*, in memory), or himself – and a player twice removed, an equity actor (Derek Jacobi in the canniest of performances, staring with barely concealed mirth at the camera), an actor playing an actor playing the prince of Elsinore feigning madness? Here, in a very real sense, the issue of performativity reaches a kind of conceptual and perceptual crisis: Which is the fiction, who is dreaming the dream of passion, or better, who most obviously telling the lie? The First player, Shakespeare, Hamlet, or Jacobi? Who, and for what purpose? "For Hecuba!" of course (echoes of Hecate, that devil, patron saint of actors), but then, "What's Hecuba to him, or he to Hecuba, / That he should weep for her?" (2.2.552–5). Where is the psychic investment? What are the expectations? Where the risk? Who, exactly, is in danger of being found out? Where is the surprise? How, exactly, is Judith Butler's idea of interpellation, the performative, running here? And then the *coup de théâtre*, the lying double bind:

> What would he do,
> Had he the motive and the cue for passion
> That I have? He would drown the stage with tears
> And cleave the general ear with horrid speech,
> Make mad the guilty and appal the free,
> Confound the ignorant, and amaze indeed
> The very faculties of eyes and ears.
> Yet I,
> A dull and muddy-mettled rascal, peak,
> Like John-a-dreams, unpregnant of my cause,
> And can say nothing.   (2.2.559–64)

"What would he do ... say nothing." Why, he would do what he (who?) is doing – act. For there is no Hamlet, and never has been – that is certainly one point of the play over time – only "this player here," the very knot and configuration of performative deceit, a deceit conceit plainly revealed, but hidden. The deceit conceit exploits the audience's desire for a "willing suspension of disbelief" and the seeming impossibility of it (both characteristics of Western theater), revealing an audience "taken" by the con: the grifter gets the mark to willfully give up his money (like the prospect of Catherine being taken from behind), and then the mark's consciousness, upon recognition, is momentarily paused – the surprise: "I've been taken" – the world has changed. But for whom? For the mark, perhaps, but certainly for the con-man/contra-man, who remains largely unconcerned with the mark's recognition of being-conned. It is of no consequence to the con-man whether the mark ever finds out. It is the game that gives pleasure – the play. And so the traditional anthem – a Hamlet "unable to act," a "John-a-dreams, unpregnant of my cause" – is itself a bald lie: that is all Hamlet can do – he acts. He is most himself, or rather most what he is, when designating himself at the outset, "this player here ..." (as Macbeth would have us believe, "That struts and frets his hour upon the stage ... Signifying nothing" [*Macbeth*, 5.5.25–8]). Hamlet tells us to our faces that he, like the first actor, *is* merely acting his passions, just "pretending," and we still do not get it because we trust him at his word, believe in his world. We are, Hamlet (or Shakespeare, or Jacobi, or someone ...) tells us to our faces, fools.

"What a rogue and peasant slave am I" (*Hamlet* 2.2.544), certainly, but also in the canny way the play surreptitiously begins over and over again, creating a *mise en abîme* of suspicion, skewed perception, and self-delusion: the opening line, "Who's there?"; Hamlet speaking to the ghost; the play within the play; the final scene, in which Horatio (echoing oration, "Flights of angels ..." [5.2.366]) gives the order:

> that these bodies
> High on a stage be placed to the view,
> And let me speak to th'yet unknowing world
> How these things came about. So shall you hear
> Of carnal, bloody, and unnatural acts,
> Of accidental judgements, casual slaughter,
> Of deaths put on by cunning and forc'd cause.
> (5.2.382–8)

He becomes the freak show barker, the porn show pimp, the stage manager ready to mount the whole bloody business one more time, right from the beginning, creating a new frame, fusing deceit and its strange attractor, repetition ("What, has this thing appeared again tonight?" [1.1.24]), into something like a genome project for theater's yet unformed history – the playing of roles whose scripts lie concealed within the fury of concealment itself, a performativity that is in reality a theater whose promptbooks and castings have been archived in the cultural unconscious, engendered in an interpellative moment that both expands and recedes hopelessly around us: "Who's there?", but also the turning back, "Nay, answer me, stand and unfold yourself" (1.1.1–2). And finally, the seeming end to it all in the invocation of the interpellation itself, "Long live the king!" (1.1.3). But which king is being hailed: the living king or the king dead?

The pith and marrow of our attribute

What is the nature of this maddening interpellative moment? In her book, *Bodies That Matter*, Butler claims that "in Althusser's notion of interpellation, it is the police who initiate the call" that establishes subjectivity (121). However, it is not only the police in Althusser's essay that initiate the interpellation, but also God, a stranger on the street, and a friend knocking at the door. It is in fact the latter that the old con-man Althusser mentions first, giving a much less authoritarian face to the process of individuation that Butler invokes, encouraging doubt about, say, one's friends, implying that our world is not as it appears, or, rather, that interpellation is not simply the result of cultural authoritarianism, but of something cellular, genetic, primal. This is no mere quibble. Butler's misreading of Althusserian misreadings is indicative of a series of misapprehensions both large and small that are in the end suggestive of critical theory's general misapprehension of apprehension itself, characterized by Butler's failure to understand the theater of her own discourse.

When Althusser introduces his critical notion of interpellation or "hailing," he gives the example of a knock on the door, followed by our question, "Who's there?" (Again, the Hamletic impasse). The friend, still unseen and unknown answers, "It's me." We open the door, and discover, indeed, that it is "her." Both she and we are constituted through reciprocal acknowledgement of presence. Similarly, out on the street someone yells, "Hey, you!" and "nine times out of ten," according to Althusser, the "correct person" responds and turns around (173–5). Putting aside the problematic question of where Althusser's "nine times out of ten" statistic comes from, his claim that this hailing or interpellation works not out of a sense of guilt – such that one says, "I turn because I know I have done something for which I might be apprehended" – but rather because the turning toward the voice, or the response, "It's me!," constitute "me" as a subject, and so in some way one comes to recognize that interpellative constitution as the means, the only means, by which one comes to exist within a society governed by "Ideological and Repressive State apparatuses" – what transversal theory refers to as state machinery.[8] The uses in either case of the universal/

personal pronouns you/me underscores Althusser's recognition, out of Jacques Lacan, that it is in the pronominal that one is constituted, not as a self with a particular name, but as a subject (of a sentence), therefore constituted in a sentence. Yet one is not necessarily sentenced to subjectivity in this process. One can resist, or even turn the interpellation into aggression: "Nay, answer me. Stand and unfold yourself" (*Hamlet* 1.1.2).

But apart from the possibility of resistance, there is also in the Althusserian interpellative moment, actually central to it, a misapprehension (*méconnaissance*) – an "otherness" to recognition that enforces its power. One might say: "When I am called, I turn because it really is me; I have not been misrecognized. At the same time, I turn to assure myself – hence the common response of pointing to oneself and asking the silently mouthed, 'Me? Really?' " Misrecognition, the surprise, the transversal movement, or its possibility, thus informs the allure of recognition and one's responses to it. Or so Althusser suggests, and Butler reiterates. Butler's contribution to the schema is to suggest that interpellation is not (as Althusser suggests, she claims) unilateral. When one is constituted as an "I," there immediately arises the possibility, as we have mentioned, that one will resist the interpellation (as Catherine may "disappear" Henry).[9]

While both Althusser and Butler are careful to point out that this process of interpellation does not work in the simple, linear, sequential way described, that the interpellative process is an "always-already," to quote Althusser, a construction or "performative" (to use Butler's misappropriation of Austin) that begins literally before we are born (the giving of a name in utero), both Althusser and Butler miss a crucial point: this process of interpellation cannot exist but for theater. Indeed, even though Butler, in her later work, has grudgingly admitted to theatrical resonances in her theory, she is adamant that the interpellative moment, the moment of performativity, is largely beyond our control, and so not, as she supposes theater to be, subject to rehearsal, but is rather reiterative, merely repetition. However, this is impossible, given that events can never be exactly repeated. She is not aware that her subjective territory, like all subjective territories, although scripted and revised in response to changes in sociopolitical conductors, state machinery, official territory, and other forces, needs to be rehearsed, adapted, and produced over and over again to be convincingly affirmed. This requires preparation, practice, interaction, and amendment in negotiation with audiences, and not just reiteration and exclusion, as Butler would have us believe, even if this were possible. To be sure, Butler does not know the secret: that transversal power can be mutually channeled through theater to endow the subject with agency to rehearse, adapt, and produce, and thereby dismiss Butler's own misuse of Austin's term "performative," to show it to be an untruth that Butler espouses. Austin's point is that language most often does something (it acts) rather than signifies something and therefore its efficacy can be seen, experienced, and measured.

Butler is not aware that consciousness occupies "theaterspace," works like theater, and vice versa. And even Althusser, for all of his invocation to theatrical metaphor in his discussion of the state apparatus, declines the theatrical consciousness suggested in the interpellated moment: "Hey, you!," re-echoing of course the Hamletic preamble, spoken by Barnardo, that sets into motion a veritable torrent of interpellation that *fails* again and again throughout the play, exposing a theater at the heart of the interpellative moment, at the moment of performativity that is predicated on the necessary failure of interpellation, a theater at the heart of performativity that has been performance, to be sure, theater, all along.

When Butler misreads this theater as the societal stage – and the stage in its most perverse form, as deceit, robotics, and evacuation – that ensures the failure of all constitutive acts, the outcome of such failure is, ultimately, not liberation (even the restricted liberation of mimicry she imagines), but rather alienation, insanity, poverty, abjection, death. The *absence* of the constituting authoritative interpellation – of framing – does not usher in possibilities for resistance, empowerment, and transformation, but rather the dissolution of identity, selves, and agency – both personal and political. Such attempts in critical theory to reduce authoritative conductors to repressive apparatuses or to negate them, not only fail, but also unproductively oversimplify what are usually complex situations: for all of their repressive tactics and modalities, authorities are often positively experienced across history as organizing mechanisms in the interest of identity formations, state power, and society, and so operate out of necessity – absolutely neither "good" nor "evil" – and thus simultaneously supply us with worlds and refuse us entry to other worlds. It is, in fact, the doubleness of authority – in the sense of its Janus-faced ability to deceive and affirm – that constitutes subjects and preconditions tragedy. It is for this reason that theater as a reconfiguring mechanism (a framed, rehearsed, developed, interactive performance praxis) can be a powerful sociopolitical conductor of transversal power self-consciously deployed in order to change the parameters that oblige us to certain authorities, and allow us to change the authorities themselves.

Earth, yield me roots

> For each true word, a blister; and each false
> Be as a cauterizing to the root o' th' tongue,
> Consuming it with speaking!
> (Shakespeare, *Timon of Athens* 5.1.131–3)

In line with a research methodology of transversal poetics commonly referred to as the "investigative-expansive mode," we would like our discussion to follow a lead. The investigative-expansive mode requires that the subject matter under investigation be divided up into variables that are then partitioned and examined in relation to other influences, both abstract and empirical, beyond the immediate vicinity. A chief objective of this expansive approach is to contextualize historically, ideologically, and critically both the subject matter and the analysis itself within local and greater milieu. Mobile and vine-like, the investigative-expansive mode resists anything resembling predetermination or circumscription. Instead, it calls for continuous analytical maneuverings and reparameterizations in response to unexpected, even sudden, emergences of glitches, quagmires, and new information as it deduces, trail blazes, follows offbeat leads, and takes tangential excursions. The present analysis encourages us to take just such a tangential foray into a related area of inquiry, primatology, as a means by which to achieve a more expansive perspective on how we are coming to understand deceit and its relationships to theater, consciousness, and transversal power.

Among other things, so far we have seen that the multifarious (or nefarious) nature of deceit is perhaps its most salient and maddening quality. The forms of deceit are so many and so varied, so seamlessly graded from one type to another, that it is difficult to determine just what the lie is, what constitutes it; and a deceit conceit is typically only recognizable in its revealed achievements, floundering, or failings vis-à-vis a designated subjective territory. Still, it may be safe to say that among the many activities that have been seen as consummately human, lying has remained within the realm of the purely social, the performed, and thus the purely human. But do other animals deceive? Of course some animals engage in mimicry (the false coral snake). Some feign injury to protect young (the killdeer). Some play dead to escape harm (the opossum). But do other, non-human animals lie on purpose; or more precisely, do animals know that they deceive; *do animals have a theory of mind?*

This remains an important question because lying has for most of our philosophical history been seen as quintessentially social, quintessentially problematic, and quintessentially a constructed, human activity. The epistemology of lying (philosophy), the ethics of lying (philosophy and theology), the analysis of lying (criminology), the uses of lying (psychology), and, finally, the hermeneutics of lying (psychoanalysis), have all been studied as primarily human behavior. Indeed, Lacan himself asked the question, "Can animals lie?" and answered "No."[10] And if animals cannot lie, they certainly cannot create illusions, delude others, employ fictions, or construct *mise en scènes*. Animals cannot, in effect, translate behaviors from one domain to another through the operations of an unconscious to produce fictive scenes, illusions, theaters. They cannot, that is, *unless they are primates* and primates of a certain level of intelligence. Primates lie, and, moreover, primates do not merely deceive using a theatricalized intelligence, intelligence in primates (and thus in humans) *is* deception. Put differently, intelligence is the combined awareness of deception's possibility and the capacity to imagine and generate creative representations.

This, then, is a true story,[11] a dumb show, not particularly subtle or ingenious, but one accompanied by, as Artaud says of the Theater of Cruelty, "cries, groans, apparitions, surprises, theatrical tricks ... a kind of unique language halfway between gesture and thought" (245; 242). A baboon, known as ML in the literary account, digs in the earth, trying to discover "one poor root."[12] A young member of the baboon troupe approaches and watches. He is known in the literature as PA. For PA time passes, but PA keeps watch, always keeps watch, yet no one else is there, no one else sees them. ML digs, PA watches, and watches to make certain others are not watching. When the root is nearly uncovered, young PA begins to scream. Hair-raising, inconsolable terror. This is what really happened, but here the story is at the same time not true – there is no danger near. No one is attacking him. It is just PA and the other baboon, ML. The troupe's alpha male, JG, hears the cry and appears over the crest of the hill. He shows raw fury, running and striking the digging baboon, who runs a short distance away, and waits for him to disappear over the crest of the hill again. When he is gone, ML resumes digging. PA, the child baboon, watches.

Now here the true story continues, that is also not true: when the root is fully unearthed, the child, PA, again begins to scream. Again, hair-raising terror. But again there is no danger. Again, no one is attacking him. Again, the protective male, JG, appears over the crest of the hill showing bared teeth. This time he runs at the digging ML and chases her away, over the crest of a different hill, and then leaves. The child-baboon, PA, now alone, picks up the cherished and much-planned-for root and begins to eat. He is, perhaps, amused, even delighted at the sensation that tickles the back of his mind. A faint sensation. Deception? Perversity? The perversity of subversion, perhaps. Or perhaps the more disturbing (for us) perversity of the non-perverse: that in the lie (though the story is true) the young baboon is following the very stage directions of a theatricalized, Darwinian mind that simply does not understand its own ... strangeness.

PA has over time used similar subterfuge. Screaming, crying, calling for help, a guardian running to help him, driving others away from food they have discovered – a scheme enabling him to eat it in their stead. Once, twice, five times – a simian con-man. And there are countless other scenarios, other actors in the troupe: different deceptions and subterfuges, some so intricate and multilayered that it is at first hard to believe they have been planned and played-out by mere juveniles and primate juveniles at that. In fact, the observations have so intrigued and puzzled primatologists, that, beginning in the late 1980s, a new theory of intelligence (and by extension, we would argue, consciousness) arose; this theory is called "Machiavellian Intelligence." Contrary to earlier assumptions – that intelligence developed in primates and subsequently in humans as a result of technical need; that being slower, weaker, lacking claws and individual killing power, primates and humans were selected for an intelligence that allowed for planning, communication, and social cooperation, and thus survival as a group – this new approach argues that intelligence was in fact selected because it allowed for the use of deception within the social sphere, and consequently enabled survival and prosperity of the individual within the group.

According to observations in the field, primatologists began to realize that primate groups, especially the more intelligent and social species (chimpanzees, bonobos, baboons), rarely utilized the sophisticated intelligence with which they were endowed for solving the logistical and "technological" problems of hunting, group-defense, or food-gathering that supposedly gave rise to higher and higher levels of intelligence. In a follow-up volume to their first book, *Machiavellian Intelligence*, Andrew Whiten and Richard Byrne explain the issue in a succinct preface:

> How can the intelligence of monkeys and apes, and the huge brain expansion that marked human evolution, be explained? In 1988,

Machiavellian Intelligence was the first book to assemble the early evidence suggesting a new answer: that the evolution of intellect was primarily driven by selection for manipulative, social expertise within groups where the most challenging problem faced by individuals was dealing with their companions.[13]

If Whiten and Byrne are correct (and the issues are controversial), the potential ramifications of this redefinition are enormous and cut across a huge range of post-humanist disciplines: primatology, of course, but studies of human development as well. These include psychology and philosophy, but also the history of consciousness, as well as studies of social behavior, ethnic and gender theory, performance theory, more precisely, and theater theory, primarily. If Whiten and Byrne's ideas hold merit, we might need to rethink central issues in psychoanalysis, Marxism, and critical theory in general: is deception, for example, and thus performance/theater, merely a constructed behavior – is it, in other words, something "grafted on," even unconsciously, to operational intelligence, or is intelligence "always already" deception? Might the perception of deception, an understanding of its etiology, be the highest, most sophisticated form of deception? Are the patterns, plans, outcomes of deception – self or otherwise – ideological? If so, is intelligence possible without ideology? Can, in other words, Marxist analysis claim the ideological high-ground if all analysis itself is predicated upon deception – strategic deceptions from the illusory "as ifs" of subjunctive space, to the densely argued discourses of propaganda?

While present modes of critical theory might presuppose a certain cynicism permeating the hegemonic forces of late capitalism, while they might, indeed, agree that deception is the rule in capital and not the exception, critical theory would almost to the person claim dispensation from the desperation of the Machiavellian double-bind, a double-bind that goes back at least as far as David Hume: that is, how can we know the truth of what we think? If deceptions, or the more creative "as ifs" and "what ifs" of subjunctivity, are the very *substance* and substrate of thought, if "the problem of representation" is not merely that it slips, or misses its object, but purposely leads us astray in a kind of perpetual gaming that is the very mechanism of our survival, then is not the rational, even in its guise as poststructural skepticism, which is to say, critical theory, nothing more than "always already" self-parody? As a discipline, critical theory, we think, would, predominantly, say no. Conveniently, this is because the gaming and illusions of thought and culture can be reapprehended, redeconstructed, the fantasy can be recast, "consciousness raised" again, ideological mis-perceptions corrected, if only partially, or "contingently." For example, in her book, *Simians, Cyborgs, and Women*, and again in *Primate Visions*, Donna Haraway describes, in the work of primatologists, something like the humanization of primate groups, or more precisely, the pro-jection of ideologies of gender onto those groups. As she puts it, "Scientific debate about monkeys, apes, and human beings, that is, about primates, is a social process of producing stories, important stories that constitute public meanings" (81). She continues:

> We find the themes of modern America reflected in detail in the bodies and lives of animals. We polish an animal mirror to look for ourselves. ... The science of non-human primates, primatology, may be a source of insight or a source of illusion. The issue rests on our skill in the construction of mirrors. (21)

While following the now unquestioned Foucaultian assertion that knowledge is constructed not discovered, even empirical, scientific knowledge, Haraway forages blindly before the subterfuge of her own intellect, an intellect that, in the terms of Machiavellian Intelligence, is born neither through a need to construct or discover, nor through a need to master an environment through technique, but rather, through the need for deceit, machination (cybernetic or otherwise), and self-delusion. Both primate and human intelligence came into being – underwent comings-to-be – because it was naturally selected for its ability to produce deceptive behavior. By extension, we are refer-ring to the lies and subterfuges, that, along with cooperative behavior, constitute social and cultural life (subterfuge cannot work, after all, unless there is trust and cooperation seemingly ahead of it). This predicament, as evidenced in Haraway's conclusions, suggests a rather profound and disturbing possibility that even the primatologists often miss: human intellect is not merely prone to dissembling, dissembling is no mere weakness within human character and mind, no aberration or "subversive" activity, dissembling and deceit are, again, what consti-tute mind – a mind constructed, as we have said, in the subjunctive modes of "as if" and "what if," in the interstices between subjective and transversal territories.

Against the accusation that such a view is overly cynical, we should remind ourselves that what is remarkable are not the sins of a species so determined, but the apparent altruism that seems to outweigh the self-interest (we have, after all, survived this long). To be sure, Machiavellian theorists would argue that self-interest, in the final analysis, is necessary to social survival at both the individual and group level: this aspect of intelligence is what allows the physically weaker (though perhaps smarter and more creative) members of the group to survive and thrive. In fact, this Machiavellian aspect of intelligence, contra Machiavelli himself, serves the purposes of the weaker individual far more than it serves the interests of the prince, or president, who everyone, in the political scheme of things, suspects is lying. And if, as Haraway might argue, the primatologists are merely seeing their own nefarious scientific practices in the actions of primates, although this is doubtful given the secretly filmed, non-edited simian sequences demonstrating deceit the primatologists provide as evidence, the idea of this projection merely begs the question all over again: Why would critical intelligence fail so easily and obviously unless in some sense consciousness wanted to be deceived – wanted to be lied to even when it knows it is being lied to (picture Derek Jacobi delivering Hamlet's "what a rogue and peasant slave am I" speech). In other words, if primatologists are projecting their own deceit onto innocent apes, what does that say about the deep structure of mind, the unconscious itself?

Critical theory, especially in its more political modalities, still seems to believe itself capable of discovering the secret ideologies of capitalism, the substrate of sexism, or the representational malaise of gender itself, while believing its own agendas to be relatively transparent. Theater, we would submit, or at least the theater that is the very embodiment of the Machiavellian mind, is not so sure. Theater, the very subjunctive location of deception, subterfuge, and creativity at polymorphous and multifarious levels, because of its ability to willfully move transversally, has always known the lying truth of the Machiavellian mind, while critical theory, in its affirmation of the possibilities of intellectual liberation, however slight, gained through endless analyses (often in bad faith) of "dominant culture," has for the most part willingly suspended disbelief in its own secret agendas and self-delusional scams. It is through the transversal power of theater, perhaps in collaboration with the critical theory that – while playing its own games – elucidates theater's opportunities, that worlds can be changed. This can happen through the transposition of deceit in theater, through deceit conceits that work simultaneously to project transversally deceit-as-other outside of subjective and official territories and to reconfigure subjunctively deceit-as-becomings. As both the mind it reflects and the praxis it performs, theater can expand our understanding and experience of deceit to include unlimited, positive potentialities in the worlds it ingeniously creates. But this, perhaps, may be impossible, if consciousness' first impulse is indeed to deceive: whereas we might believe we are working for enlightenment and liberation, we may be, unbeknownst to us, seeking our own desecrations and death. What do you tell yourself?

Notes

1. For more on "subjunctive space" beyond coverage in this book, see Bryan Reynolds, "Transversal Performance: Shakespace, the September 11 Attacks, and the Critical Future" in his *Performing Transversally: Reimagining Shakespeare and the Critical Future* (New York: Palgrave Macmillan, 2003): 1–28. For detailed discussion on "subjective territory" and "transversal territory" beyond coverage in this book, see Bryan Reynolds: "The Devil's House, 'or worse': Transversal Power and Antitheatrical Discourse in Early Modern England" (*Theatre Journal* 49.2 [1997]: 143–67); *Becoming Criminal: Transversal Performance and Cultural Dissidence in Early Modern England* (Baltimore: Johns Hopkins University Press, 2002), 1–22; and *Performing Transversally: Reimagining Shakespeare and the Critical Future* (New York: Palgrave Macmillan, 2003), 1–28.
2. But we need to ask, of course, if this is in fact how "performance" or the performative is currently understood. Rather, out of Butler et al., performance seems precisely that mode of "doing" that *lacks* self-consciousness. "Performance," as the term is currently employed, seems to mean little more than "doing." See in this regard Jon McKenzie, *Perform or Else: From Discipline to Performance* (New York: Routledge Press, 2001).
3. The term "pressurized belongings" was coined by Glenn Odom and Bryan Reynolds in their essay, "Becomings Roman/Comings-to-be Villain: Pressurized Belongings and the Coding of Ethnicity, Religion, and Nationality in Peele & Shakespeare's *Titus Andronicus*," which is Chapter 8 of this book.
4. On Shakespace, see Donald Hedrick and Bryan Reynolds, "Shakespace and Transversal Power," in Hedrick and Reynolds Ed., *Shakespeare Without Class*: 3–50; and for more on articulatory spaces, see Reynolds, *Performing Transversally*: 1–28.
5. See Stephen Blaha, *Cosmos and Consciousness: Quantum Computers, Super strings, Programming, Egypt, Quarks, Mind Body Problem, and Turing Machines* (Auburn, New Hampshire: Pingree-Hill Publishing, 2002) and Daniel Dennett, *Consciousness Explained* (Boston: Little, Brown and Company, 1991).
6. See Jacques Lacan, "Desire and the Interpretation of Desire in *Hamlet*," *Literature and Psychoanalysis: the Question of Reading: Otherwise*, ed. Shoshana Felman (Baltimore: Johns Hopkins University Press, 1982).
7. Herbert Blau, *The Dubious Spectacle: Extremities of Theater, 1976–2000* (Minneapolis: University of Minnesota Press, 2002): 107.
8. Reynolds coined the term state machinery for a society's governmental assembly of conductors as a corrective to the political philosophy of

Althusser. With state machinery, a term that simultaneously connotes singularity and plurality, he adapted Althusser's conception of what he calls the "Repressive State Apparatus," which includes the governmental mechanisms, such as the military and the police, that strive to control our bodies, and fused it with his subsidiary "Ideological State Apparatuses," the inculcating mechanisms that strive to control our thoughts and emotions. Reynolds emphasizes that a society's drive for governmental coherence is always motivated by assorted conductors of state-oriented organizational power that are at different times and to varying degrees always both repressive and ideological. This is a sociopower dynamic in which various conductors work, sometimes individually and sometimes in conjunction with other conductors, to substantiate their own positions of power within the sociopolitical field. Hence, use of the term state machinery should make explicit the multifarious and discursive nature of state power, and thus prevent the misperception of the sociopower dynamic as the result of a conspiracy led by a monolithic state. This is not to say, however, that conspiracies do not occur and take the form of state factions. On the contrary, this must be the case for the more complex machinery to run. See Reynolds, *Becoming Criminal*, 1–22.

9. The very process that constitutes a "me," therefore, is the means through which an "I" can challenge that constitution. Dick Hebdige perhaps says it best when discussing the punk movements of the seventies: the reason for the piercing, provocatively torn clothing and scarification of punk teenagers living within the surveillances of the British police and social service agencies was designed to "make being looked at an aggressive act." The quote is from a talk Dick Hebdige gave at the University of Wisconsin in 1985 at the Center for 20th Century Studies.

10. See, for example, Lacan's discussion of signification in the process of analysis in *Ecrits: a Selection*, translated by Bruce Fink (New York: Norton, 2002), 88–90.

11. Perhaps this is the central concern of this essay, the very assertion "this is a true story": what, exactly, does truth amount to in fiction, in narrative, in theater? How can we believe anything, even our capacity to lie, given that deceit is the ground condition of consciousness? Even more disturbing, how can you (the reader) believe what you are reading at this very moment? What are *your* (whose?) secret agendas, self-delusions, con-games? How can you, finally, even believe in the presence of deceit, if deceit is the condition of consciousness? What, in other words, we see in the primates' lies and deceits is being recapitulated at this very moment, in this very essay. What are we not telling you? What are we leaving out of the story? What theoretical niceties are we invoking, not to elucidate, but to, in essence, cover-up what we do not want you (or ourselves) to know or think about? Paranoia, or simply fear, of "being on the outside, being politically misaligned, not getting tenure ..."

12. The following baboon story is taken from Richard Byrne and Andrew Whiten, *Machiavellian Intelligence: Social Expertise and the Evolution of Intellect in Monkey, Apes, and Humans* (Oxford: Clarendon Press, 1988): chapter 15.

13. Byrne and Whiten, *Machiavellian Intelligence*: front material.

Works cited

Althusser, Louis. "Ideology and the State." *Lenin and Philosophy*. New York: Monthly Review Press, 1971.

Artaud, Antonin. "The Theater of Cruelty: First Manifesto." *Antonin Artaud: Selected Writing*. Ed. and introduction Susan Sontag. New York: Farrar, Straus and Giroux, 1976.

Blaha, Stephen. *Cosmos and Consciousness: Quantum Computers, Super strings, Programming, Egypt, Quarks, Mind Body Problem, and Turing Machines*. Auburn, New Hampshire: Pingree-Hill Publishing, 2002.

Blau, Herbert. *The Dubious Spectacle: Extremities of Theater, 1976–2000*. Minneapolis: University of Minnesota Press, 2002.

Byrne, Richard and Andrew Whiten. "Tactical deception of familiar individuals in baboons." *Machiavellian Intelligence: Social Expertise and the Evolution of Intellect in Monkeys, Apes, and Humans*. Ed. Richard Byrne and Andrew Whiten. Oxford: Clarendon Press, 1988.

Butler, Judith. *Bodies That Matter: On the Discursive Limits of "Sex."* New York: Routledge: New York, 1993.

Dennett, Daniel. *Consciousness Explained*. Boston: Little, Brown and Company, 1991.

Haraway, Donna. *Primate Visions: Gender, Race, and Nature in the World of Modern Science*. New York: Routledge, 1989.

—— *Simians, Cyborgs, and Women*. New York: Routledge, 1991.

Hedrick, Donald and Bryan Reynolds. *Machiavellian Intelligence II: Extensions and Evaluations*. Edited by Andrew Whiten and Richard Byrne. Cambridge: Cambridge University Press, 1997.

—— " 'A little touch of Harry in the Night': Translucency and Projective Transversality in the Sexual and National Politics of *Henry V*," in Bryan Reynolds, *Performing Transversally: Reimagining Shakespeare and the Critical Future*. Palgrave Macmillan: New York, 2003: 171–88.

—— "Shakespace and Transversal Power," in Hedrick and Reynolds Ed., *Shakespeare Without Class: Misappropriations of Cultural Capital* (New York: Palgrave Macmillan, 2000): 3–50

Reynolds, Bryan. *Performing Transversally: Reimagining Shakespeare and the Critical Future*. Palgrave Macmillan: New York, 2003.

Shakespeare, William. *King Henry V*, Ed. Andrew Gurr. Cambridge and New York: Cambridge University Press, 1992.

—— *Macbeth*. Ed. Kenneth Muir. London: Methuen, 1951.

—— *Hamlet*. The Arden Shakespeare. Ed. Harold Jenkins. New York: Routledge, 1989, 1990.

—— *Timon of Athens* The Arden Shakespeare. Ed. H. J. Oliver. New York. Methuen, 1959.

4
Comedic Law: Projective Transversality, Deceit Conceits, and the Conjuring of *Macbeth* and *Doctor Faustus* in Jonson's *The Devil is an Ass*

Amy Cook & Bryan Reynolds

> If [Satan] be not / At all, why are there conjurers?
> If they be not, / Why are there laws against 'em?
> *(The Devil is an Ass* 1.2.20–2)

Early modern England's state machinery and official culture used the Holy Bible as the legal basis and rationale for its ordering procedures. To promote the populace's "appropriate" subjectification, the dominant sociopolitical conductors relied most of all on the – perhaps overdetermined – "objective agency" of this traditionally consecrated historical artifact as its transcendental signified.[1] Objective agency is the power of a sign or sign-object (the latter of which applies here) to frustrate or transcend the effectual parameters and expectations generally ascribed to that sign by a system of codification within which it functions. Because the Bible was believed to be the word of God, only comprehensible in the final analysis by God himself, the Bible's meaning and agency, always *in potentia*, as prescribed by the all-knowing creator, could not be infallibly determined by imperfect humans.

The Bible achieved, as a result of its popular fetishization (perhaps as a meme machine, as described in Chapter 1), fantastic and powerful "affective presence" – the combined material, symbolic, and imaginary existence of a concept/object/event – across spacetime,[2] helping to transpose, in the minds of the populace, early modern monarchs into divine rulers, barmaids into succubae, and the public theater into the institution of Satan, what the period's antitheatrical moralists called "Sathans Synagogue" (Heywood 43).[3] However negotiated through translation, exegesis, and edition, the Bible resolutely affirmed for early modern

England's state machinery that God neatly divides human from animal, man from woman, King from commoner, good from evil, and true from false. In contrast, it maintained that Satan, as the source of evil in God's dominion, is the metaphysical "fugitive element" (an authority-under-mining agent, which often generates transversal power) of defiance and corruption working to confuse and problematize these distinctions, to cause the violation and destruction of the moral and spiritual bound-aries separating and distinguishing them. The official conception of God-as-source-of-order and Satan-as-source-of-confusion is epitomized, for example, in the homily "Exhortacion concernyng Good Ordre and Obedience to Rulers and Magistrates" (first printed in 1547, and reprinted regularly during Elizabeth's reign) that the state generated for preaching in Church services, where weekly attendance by the populace was mandatory (although apparently not strictly enforced):

> Almightie God hath created and appointed all thinges in heaven, yearth and waters in a moste excellent and perfect ordre. ... Some are in high degre, some in lowe, some kynges and princes, some inferiors and subjects, priestes and laimen, masters and servauntes, fathers and chyldren, husbandes and wifes, riche and poore, and every one hath nede of other. ... Where there is no right ordre, there [Satan] reigneth all abuse, carnall libertie, enormitie, syn and babilonicall confusion. (Bond 161)

Like fugitive elements in any system, Satan, like the witches in *Macbeth*, the too-curious conjuring Doctor Faustus, and the devil Pug in *The Devil is an Ass*, eludes capture and stasis, undermines authority (the "right ordre"), and resists all delimiting, reducing, and totalizing machinations. The Bible, as the ultimate textual authority for the legal system in early modern England, defined the terms of the debate around ordered binaries and worked to ground the fugitive, just as it continues to influence anti-fugitivity and have a profound impact on the sociopolitics of twenty-first century life, particularly in the West.

Ben Jonson's *The Devil is an Ass* (1616) challenges the Biblical order by suggesting that Satan, as God's proxy for the manifestation of disorder, is not effectively fugitive because he cannot successfully affect humans, as this would make humans ultimately solely responsible for their iniquity and, implicitly, their goodness as well, in what appears to be a closed system (or cosmos). In line with transversal poetics, we will pursue this contention within *The Devil is an Ass* using the "investigative-expansive mode of analysis" (i.e. mode) that first dismantles the subject matter under investigation into its constituent parts and then relates them to other forces, both abstract and empirical, extending beyond the immediate parameters. Contextualizing historically, ideologically, and critically the subject matter and the analysis itself within various environments across spacetime, this mobile approach continually reparameterizes in response to the unexpected emergence of glitches and new information, and resists any predetermination or circumscription. The employment of the i.e. mode allows us to explore, via early modern English Satanic discourse, modern theories of comedy, *Doctor Faustus*, and *Macbeth*, what appears to be Jonson's suggestion that humans's complications are often generated from within their subjective territory through doubt and confusion engendered by an idea of Satan, but not by Satan himself, thus rendering humans with greater knowledge and therefore greater power than the Biblical system would tolerate. We want to show that Jonson's play in fact offers an alternative law to that of the Bible, functioning similarly to equivocation in *Macbeth*, which shines a critical light on early modern England's dominant theology-ideology by proposing that Satan actually has little or nothing to do with people's immorality and sinfulness. Instead of reinforcing belief in Satan's evil interventions, the play seems to maintain that people, without influence from Satan or anything metaphysical or supernatural, are responsible for their own actions and/or the actions of others, whether deemed bad or good by the communities within which they live and operate. Nevertheless, the play also seems to maintain that Satan and his minions (lesser devils and vice figures) are active forces in both the physical world and hell, capable of intermingling and interacting with humans at will. But how can Jonson have it both ways?

Satanic machinations: pains in the ass

The answer to this question may lie in Jonson's agreement with a then widely circulated view of Satan as psychological infiltrator and manipulator of people rather than demonic possessor of people's bodies, beingness, and souls. At the beginning of *The Devil is an Ass*, when the lesser devil Pug requests permission from Satan to intervene mischievously with the goings-on of humans on earth, Satan responds mockingly with a litany of references to both characteristics common to witchcraft allegations and to real-life cases of alleged witchcraft and *maleficia* that were topical and controversial for much of Jonson's audience (note our annotations in brackets within the quotation):[4]

You have some plot now / Upon a tunning of ale, to stale the yeast, / ... Or some good ribibe [old woman] about Kentish Town, / Or Hoxton, you would hang now for a witch, / Because she will not let you play round Robin? [a children's game, since women were often convicted of witchcraft on the testimony of a single child] / And you'll go sour the citizens' cream 'gainst Sunday: / That she may be accused for't, and condemned / By a Middlesex jury [which was notorious for being corrupt], to the satisfaction / Of their offended friends, the Londoners' wives, / Whose teeth were set on edge with it? ... / You would make, I think, / An agent to be sent, for Lancashire / Proper enough [because Lancaster's court was notorious for hanging witches]. (1.1.12–33)

Satan's understanding of the workings of his minions is consistent with a perspective disinterred by Jonathan Gil Harris: "Rather than dispute the biblical proscription against witches and witchcraft, Tudor and Stuart authorities attempted to resolve the impasse simply by displacing the cause of malevolent transformation from the tongue [words spoken] to an external, infiltrating agency: Satan" (124). The power of the charms and curses of witches was denied, and this denial was consistent with the Protestant's dismissal of Catholics' investments in the magical power of words. Put differently, according to the Protestants, because witches and Jesuits erroneously and blasphemously assigned powers to words, they were more susceptible to Satan's machinations.

In his *Dialogue Concerning Witches and Witchcraftes* (1593), George Gifford argues, "It is to be accounted among the vile and filthy abominations of popery, that they ascribe a power to driue away Deuils vnto words & sillables pronounced ... for when as they ascribe power vnto such things to driue out deuils, what are they but Witches" (qtd. in Harris 123). In order to resolve the problem of how to dismiss the verbal powers of Catholics and witches and still claim them as Satan's workers, "theologians were inclined to regard the efficacy of the witch's charms as illusory, but they nonetheless conceded a second-order effectiveness which they attributed to the covert assistance of the devil" (Harris 124). In *Mystery of Witchcraft: Discouering, the Truth, Nature, Occasions, Growth and Power Therof* (1617), Thomas Cooper insists: "Satan ... perswades the *Witch*, that whatsoeuer euill ensues, proceeds from the virtue of that curse, and not from his secret helpe" (qtd. in Harris 124). The views of these writers corroborate Clifford's, which Harris eloquently summarizes:

The devil is not capable of miracles; but his sophisticated understanding of the laws of God's creation enables him to know things
beyond the grasp of humans. For example, he is able to recognize
those people suffering, unbeknownst to themselves or others, from
a terminal illness, and to foresee the approximate date of their
death. He can easily contrive, therefore, to engineer a squabble
between a dying person and a witch; she curses that person, and
upon their inevitable demise shortly thereafter, she is persuaded
that her curse was responsible. (124)

We want to take the implications of the discourse displacing responsibility, but not liability, from the witches onto Satan to consider the
questionable logic behind the assertions made by these writers.
According to them, however subtextually expressed, it appears that
Satan's real interventions were in the creation of circumstances precipitating the unjustified persecution of witches for actual evil deeds,
while the condemnation of witches for conjuring was nevertheless still
justified. As Dennis Kezar notes, "When these transactions are viewed
from the social level, the community behaves like a possessed body
and the witch like the threatening demon; in order for the community
to regain possession of itself, it must effectively exorcise the witch"
(127).[5] Unable to directly influence the natural world, say, by causing
illness and death, Satan instead promoted terror and hysteria that
effectively subverted the ordering principles of official society in early
modern England. Satan's psychological terrorism fractured minds,
families, and communities; it inspired controversially fanaticism and
skepticism over both the Biblically supported anti-witchcraft laws – as
informed by Exodus 22:18 ("Thou shalt not suffer a witch to live") –
and the people officially endowed with the power and job of enforcing
them. In *The Devil is an Ass*, Jonson uses the period's witch craze to
reorder the logic of terrorism inspired by the witch trials[6] to expose
positive aspects of doubt and confusion.

Comedic law and orderlessness

Jonson's play demonstrates that doubt and confusion can generate
comedy. This "comedic law," that we propose supersedes the Biblical
law in Jonson's play, works to endow an audience of subjects with a
sense of humor to find meaning within what transversal theory refers
to as "deceit conceits." Deceit conceits are clever schemes involving
artifice and fiction performed in order to fracture, transform, and/or

expand the conceptual-emotional range of an individual or individuals, in other words, a targeted subjective territory. The ability to think beyond the limits of this territory is crucial to comedy since jokes generally require a breaking away from norms and an excursion into the realm of what transversal theory refers to as "subjunctivity," the hypothetical, "subjunctive spaces" of "what ifs" and "as ifs" that usually make thought, emotion, and experience outside of the familiar (ostensibly) possible.[7]

During the Medieval and early modern periods this was done sometimes through what scholastics called *secundum imaginationem* ("according to the imagination"), which was a concept that allowed people – similar to the Jesuits' use of equivocation (see Chapter 1) – to move subjunctively without committing heresy; people could claim that they are simply "imagining" alternative theologies while simultaneously not believing in them, and therefore not challenging the only legitimate source for metaphysical truth: the Bible.[8] To the extent that it can be applied to most comedic situations, comedic law states that thinking beyond the (ostensibly) possible questions what is considered to be possible (in the case at hand, according to God's law). Yet, according to the writers of tracts on witchcraft already discussed and to the period's antitheatrical moralists, it is precisely within the bold, mysterious, and unstable environments of subjunctive spaces, where the impossible is considered possible, that thought is potentially ungodly, where freewill allows for the will of Satan to intervene, that God's law and, by extension, the official law of early modern English society are most powerfully suspended and interrogated, if not rendered flawed and/or arbitrary.

The antitheatricalists maintained that the public theater, because it is an institution of Satan, works to both demonically possess people and project people outside of their prescribed subjective territories, moving them transversally into becoming sinful. In *Playes Confuted in Five Actions* (1582), Stephen Gosson argues that, "*Stage Plaies are not to be suffred in a Christian comon weale*" because they "are the doctrine and inuention of the Deuill" (B3). This view is supported by John Greene in *A Refutation of the Apology for Actors* (1615) in which he gives a striking account of "a Christian woman [who] went into the Theater to behold the plaies":

She entered in well and sound, but she returned and came forth possessed of the Diuell. Wherevpon certaine Godly brethren demanded Sathan how he durst be so bould, as to enter into her a

Christian. Whereto he answered, that *hee found her in his owne house*, and therefore took possession of her as his own. (44)

Early modern antitheatricalists ascribe to the theater, as an open terrain where the imagination runs wild, a devilish power to alter transgressively the minds and wills of men and to foster demonic possession.

Consistent with the antitheatricalist perspective, *The Devil is an Ass*'s comedic law encourages the audience to consider certain "what ifs" and "as ifs," to enter a subjunctive space that contests God's law. The overall effect of this draw is thus the play's invocation of an investigation *secundum imaginationem*. *The Devil is an Ass* questions the antitheatricalists' criticisms of certain performance conventions and strategies of the public theater, especially what transversal theory terms the theater's "projective transversality."[9] Projective transversality accounts for the machinations by which one spurs transversality in the other rather than in oneself, an orchestrated moving of an other outside of their established subjective territory. *The Devil is an Ass* proposes that the artifice of the theater, particularly its projective transversality, has been adopted by, or finds its source in, early modern England's opportunistic mercantile class. The play suggests that its audience, infatuated with the power of deception and performance, fails to accurately read through fiction, equivocation, or rhetoric to assess the truth masked within.

The Devil is an Ass works, as we hope to show, to critique the Biblical grounding on which the antitheatricalists's arguments are based. The play suggests, especially through the character of the devil Pug, that it is not just in the theater where identity transformations are enacted, but by Londoners at large desperately trying to become others. The use of deceit conceits and enacted projective transversality is very different from the impersonations that characterized the public theater. In Jonson's play, the characters endeavor to become imagined others alternative to their scripted realities. The guiding principles behind the projections ventured by the characters involve intangible and material aspirations, with socioeconomic positioning, measured by prestige, taking precedent. In the process, as *The Devil is an Ass* asks us to question the Biblically-grounded laws as well as the existence of Satan, which throws into relief the ideology that requires Satan for substantiation, it pulls a switcheroo: the play discloses, with masterful irony, that the play itself is in fact a product of the Satan imagined by the antitheatricalists, and that Jonson himself, having been infiltrated *secundum imaginationem* (psychologically), has presented us with an exemplary equivocation by passing off Satanism as antitheatricalism.

The Devil is a joke: comedic lines of flight

In comedy, otherwise disparate worlds, words, and concepts can intermingle and flee to cause the ephemeral juxtaposition or false marriage that the listener recognizes as funny. Comedy can confuse and create doubt. It can reorder the existing logic and expose the possibility of alternate ways of thinking. These potential alternatives might not fit the simple, ordered world of the dominant paradigm and may allow us to imagine new worlds. Hence, comedy has the potential to facilitate transversal movements across subjective territories, spreading doubt like the Devil, and can reward comedic and, by extension, conceptual departure with the generation of laughter. Before moving on with our discussion of *The Devil is an Ass*, in a move characteristic of the investigative-expansive mode of analysis, we will offer a brief inventory of important twentieth-century theories of comedy that work to productively situate our conception of comedic law in relation to a common denominator to comedy that seems to illuminate Jonson's enterprise: however short-lived, humor typically involves a departure from ordinary thinking.

For Mikhail Bakhtin, successful comedy, specifically in the form of the carnivalesque, necessitates inversion and derision of established norms.[10] Henri Bergson sees jokes as the unseating of the moral by the physical. Seen from a distance, he argues, the replacement of the neat and rational by the chaotic and corporal is funny.[11] Sigmund Freud argues that all jokes are subversive in that they demand the temporary overthrow of the conscious in favor of the unconscious. He argues that the pleasure brought about in the listener is due to the way the joke "will evade restrictions and open sources of pleasure that have become inaccessible" (103). Exposing and puncturing previously inviolable spaces, jokes release pent up energy and repression; "the joke-work makes use of deviations from normal thinking" (60). Mary Douglas subscribes to Freud's basic conception, but explores the role of the context and social situation in examining why certain situations call for this treason of reason.[12] Neuroscientist V. S. Ramachandran sees humor as an evolutionary reaction to a shift in the perceived danger of a situation. Cognitive linguist Seana Coulson argues that comedy results from the semantic leap necessary to bridge two disjointed cognitive frames. From Bakhtin to Coulson, the twentieth century has seen comedy become more and more the purview of the ordering attempts of scientific thinking. To trace this thinking and how it might illuminate Jonson's comedy, we will look more closely at the work of

Ramachandran and Coulson. Following its fugitivity, we read *The Devil is an Ass* through the sense of humor it encourages and the pleasant shaking of mind and body caused by the resulting laughter.

Ramachandran sees an evolutionary role of laughter: the release of a laugh communicates to the group that what may be perceived as an anomaly or threat is actually nothing to worry about, thus allowing for the social-acceptance of advantageous mutations that might otherwise be extinguished. He connects the teeth-baring expression of aggression with the smile, and argues that a smile is an aborted aggressive act. In other words, when the situation that once seemed to require hostility is suddenly perceived differently, the face changes to reflect, and communicate, the recognition of this change. When someone describes a situation that at first seems one way and is then revealed in another compatible but surprising light, the listener recognizes a joke and laughs. The listener must evaluate the same data from at least two different perspectives, and this paradigm shift can create confusion in the realignment of logic necessary to process the information given. The laughter that results communicates that the listener understands and has followed a comedic line of flight. Moreover, Ramachandran describes laughter as the way "you distract yourself from your anxiety by setting off your own false alarm mechanism" (207). Setting off this false alarm mechanism, then, allows a society to "playfully juxtapose larger ideas or concepts – that is, to be creative. This capacity for seeing familiar ideas from novel vantage points (an essential element of humor) could be an antidote to conservative thinking and a catalyst to creativity" (206).

Ramachandran's renderings of comedy find a bedfellow in Coulson's theories. Coulson outlines a theory of meaning construction that involves accessing different frames of reference that are then switched and juxtaposed to create meaning in the interstices. Frame shifting, she argues, is what happens when one listens to a joke, as information arises that is unexpected and forces us to reevaluate the background information and expectations operating tacitly under our comprehension: "Everyone had so much fun diving from the tree into the swimming pool we decided to put in a little water" (34). When we read "water" at the end of that sentence we realize that the frame we had retrieved for "swimming pool" included a slot for "containing water," and therefore we must shift frames to accommodate this different situation. The information that the pool was empty while everyone had fun diving from the tree requires different background information that will take into account the kinds of people who enjoy this activity.

Frame shifting "reflects the operation of a semantic reanalysis process that reorganizes existing information into a new frame" (Coulson 34). A joke requires this reorganizing; it sows doubt in our current frame structure as frame layers must combine and shift to accommodate the information provided in the joke.

Both Coulson and Ramachandran see laughter and comedy as operating within the liminal space between two cognitive states – in a subjunctive space. A situation that escapes preliminary categorization or merges seemingly opposing truth-claims can defy the assumptions and cognitive boundaries of subjective territory generated and maintained by the state machinery of a society. Such a situation was identified by early modern England's antitheatricalists, who were not interested in, as Ramachandran puts it, "an antidote to conservative thinking" (206); instead, they viewed the blurring of perceived truths caused by theater, comedy, and subjunctivity as the Devil's work.

Such an example of the conflicting and confounding of "realities" that would have been threatening to the antitheatricalists can be found in the humor in the scene with the Spanish Lady in *The Devil is an Ass*. These "realities" included: 1) Fitzdottrel thinks the Spanish Lady speaks expertly about being a courtly lady; 2) the Spanish Lady is neither Spanish nor a lady, but rather Wittipol who is using the disguise to gain access to Fitzdottrel's wife; and 3) the actor playing Wittipol, playing the Spanish Lady, is very tall and not a typical boy player. Each perceived truth must be read in relationship to the others, and thus is reliant upon the acknowledgment of the others. For instance, Jonson has reminded the audience prior to this scene that the actor playing Wittipol (playing the Spanish Lady) is the actor Dick Robinson, an actor noticeably taller than the other boy players. When discussing who they could get to play the Spanish Lady, Merecraft and Engine decide on Dick Robinson, and when Wittipol shows up as the Spanish Lady, Merecraft worries that he is too tall. Engine responds: "Robinson's as tall as he" (3.4.14). To get the joke, the audience must engage *secundum imaginationem*, entertaining a situation in which one person can be many things at once: actor, character, man, and woman. Situations, meanings, and identities can slip out of their stable moorings and become something other when viewed from a different perspective. The doubt caused by this temporary confusion requires a rethinking and reordering which destabilizes other assumptions and perspectives. The joke that accomplishes this disrespects the divisions of God's ordered universe, calling into question assumptions that the early modern state machinery used to define and interpolate its subjects. As evident in the Spanish Lady scene, *The Devil is an Ass* uses comedy to trouble the simple dialectical constructs order/confusion, male/female, and truth/performance. Noted for the antitheatrical tropes within his texts,[13] Jonson in fact uses comedy to perform the same subversive act of confusion employed by the Devil. As Jonson's title suggests, the Devil is a joke.

Jonson's judge and jury: deceit conceits in the theater and courtroom

The Devil is an Ass traces one elaborate con after the other written and directed by Merecraft, who is listed in the play's dramatis personae as "The Projector." The first citation for projector in the Oxford English Dictionary is from 1596 and defines it as, "one who forms a project, who plans or designs some enterprise or undertaking; a founder." Like Merecraft, the founder projects ideas, others, and enterprises into a future site of their manifestation. From a foundational story, the projector moves his audience subjunctively into a newly-imagined space. He performs a projective transversal act in providing access to conceptual and emotional space previously unavailable, and, by extension, to the corresponding physical spaces as well. Merecraft's project, to con Fitzdottrel out of his money, succeeds on the basis of the fictions he creates. The idea that a project or enterprise begins in the thoughts and discourse of people and then can be actualized socially suggests that thinking is necessary to becoming, and that for people to become something different from what they are, their new identity must be performed convincingly for the audience, and thus comprehended and contemplated by others. Jonson's comedy exposes the limits of discourse and places the responsibility for truth at the feet of the judge and jury watching the performance.

Merecraft, like Shakespeare's projector Prospero,[14] orchestrates deceit conceits that consume the participants and manifest the alternate reality of his plans. The Epilogue of *The Devil is an Ass* – "Thus the projector here is overthrown. / But I have now a project of mine own" (1–2) – echoes and rhymes the Epilogue of Shakespeare's *The Tempest*: "Now my charms are all o'erthrown, / And what strength I have's mine own" (1–2). In both plays, the projector may be overthrown, but the project continues, however short-lived. Whereas Shakespeare, it has been argued, retires himself with his projector, Jonson's Epilogue suggests that the playwright takes over for the projector: "If it may pass: that no man would invite / The poet from us to sup forth tonight / If the play please. If it displeaseth be, / We do presume that no man will – nor we" (3–6). Jonson's project involves assessing the success of his play's performance based on whether or not the poet is taken to dinner. This bookends the discussion begun in the prologue about the role of the audience as spectator, participant, and judge of the play. Through depicting a courtroom scene in his play in which the judge's role as audience is conflated with the audience's role as judge of the play as a whole, Jonson indicts the reception of the performance by the audience as judge, rather than the fictions within the performance.

A projector uses a fiction, a deceit conceit, to create a desired and unconventional truth, but the project, in the planning stages, is nothing more than a story. It is only when key participants engage and believe the story that it can become reality. Just as Satan's work is not in the possessing of witches but in supporting the story of witchcraft in communities, the doubt and confusion spread by Satan, a joke, or a projector has the power to generate new truths in the faultlines created by the cognitive tremors needed to comprehend the fiction. In *The Belman of London*, Dekker describes the elaborate criminal cons played by a "puncke" (prostitute) and her partner, the "ruffian" (common criminal). Once the puncke has coyly lured a man to drink with her, the ruffian enters as her husband to defend her honor: "in comes a Ruffian with a drawne rapier, calles the Puncke (as she is) damned whore, askes what Rogue that is, and what he does with his wife. The conclusion of all this counterfeit swaggering being a plot betwixt this panderly ruffian and the whore to geld the silly foole of all the money hee hath in his purse" (Dekker 144). The puncke's con is similar to Merecraft's in that she wants to project a narrative of seduction for the unsuspecting victim and from within this alternate space to take his money. Dekker calls these cons "comedies" (144), describes how they are played within the "Sacking-Law," and lists the cast of characters.[15] Jonson has Merecraft play this type of comedy on Fitzdottrel, exposing the powerlessness of an actual devil to compete with a projector's devilish tricks. Jonson uses a comedic departure, which both is fugitive and spawns fugitive elements, to complicate his audience's simple reception of the performance.

In *The Devil is an Ass*, official law and the social codes that define it are also complicated insofar as they are constantly undermined and unmasked as unstable tactics of communication and control. Referring to the Devil, Fitzdottrel ingenuously argues for the potential projective transversality of official religious and legal discourse: "If he be not / At all, why are there conjurers? If they be not, / Why are there laws against 'em?" (1.2.20–2). In line with his officially-prescribed subjective territory, Fitzdottrel believes in the Devil because the narrative scripted by the law affirms that conjurers are capable of conjuring the Devil. However, it becomes clear that the story generated by the law can conjure more successfully than the conjurers themselves. A few scenes later, Jonson uses Wittipol's seduction of Fitzdottrel's wife to establish the manipulability of social codes rhetorically engaged and/or contextualized narratively to conceive and deceive. Wittipol offers Fitzdottrel an expensive cloak which Fitzdottrel hopes to wear to the Blackfriars Playhouse where he will "Publish a handsome man, and a rich suit" (1.6.34) in exchange for twenty minutes to woo Mistress Fitzdottrel. Fitzdottrel accepts the trade, assuming that Wittipol can do little with words: "Use all the tropes / And schemes that Prince Quintilian can afford you: / And much good do your rhetoric's heart" (1.4.100). The audience, likely to recognize that Fitzdottrel is a fool, would perceive his line ironically, and thus anticipate a full rhetorical assault on Mistress Fitzdottrel. Even if Fitzdottrel has not read Quintilian, who he sarcastically promotes to "Prince," Jonson certainly has.

A classical rhetorician who was taught in grammar schools in the sixteenth-century, Quintilian argued for the power of emotionally-charged visions to persuade an audience:

> When the mind is unoccupied or is absorbed by fantastic hopes or daydreams, we are haunted by these visions of which I am speaking to such an extent that we imagine that we are traveling abroad, crossing the sea, fighting, addressing the people, or enjoying the use of wealth that we do not actually possess, and seem to ourselves not to be dreaming but acting [read "doing," not "performing"]. Surely, then it may be possible to turn this form of hallucination to some profit. (6.2.30)

The hallucinations created by the projector, in this case Wittipol, can generate the same emotions as the act of doing in the listener, in this case Mistress Fitzdottrel. Wittipol begins by telling Mistress Fitzdottrel that he shall not use "false arts" or "charms" on her (1.6.108), but nonetheless proceeds to conjure her agreement through his words, further exposing the constructedness and plasticity of conventions. When she does not respond to his proposal, he responds for her, becoming her in a projective transversality that creates the submission he articulates. She does not speak until Wittipol has left, but when she does, the form and content of her lines seem to echo what she heard from Wittipol, as if by speaking for her Wittipol begins to speak within her. She refers to Wittipol as a "young gentlemen," implying that she did, as he requested, "examine both my fashion, and my years" (1.6.124). When she then refers to her room as her "cage" (1.6.238), Jonson establishes that Wittipol's story of Fitzdottrel as her "gaoler" has successfully shifted her perception of her own reality. Wittipol and Merecraft, savvy poets and performers, can rely on their onstage audience, unaware of the multiple performance frames that the offstage audience would be subject to, to fall for their hallucinatory fictions without taking into consideration the performance conditions that inform the narrative.

The same mutability of established structures, conceptual and material, that makes a con a comedy also makes a courtroom comedic. Cuckolded by Wittipol's words, Fitzdottrel uses his rhetorical wherewithal to con Justice Eitherside in the courtroom scene. Once Fitzdottrel has fallen prey to the deceit conceits of all the other characters, he uses what he has learned to con the court. Under Merecraft's tutelage, Fitzdottrel theatrically performs his possession by the Devil. Though his foaming mouth and swelling belly add to the scene, Justice Eitherside seems most convinced by the carefully chosen words Fitzdottrel uses. When Fitzdottrel repeats, "Yellow, yellow, yellow, yellow" (5.8.74), Justice Eitherside immediately reads it as a reference to Anne Turner's yellow starched collars: "That's starch! The Devil's idol of that color" (5.8.75). Anne Turner, who was hung for her alleged involvement in Sir Thomas Overbury's murder, wore the yellow starched collars that she had made fashionable to her execution.[16] Jonson invokes the color to conjure the famous witch in the minds of the audience. Fitzdottrel's interjections are all codes meant to refer to the commonly held conceits about possession. He claims that his wife pricks him, he repeats "yellow," he rhymes, and speaks several languages. He does not state what is happening to him or even "act" as much as Merecraft would like him to. He does not have to, for by referring to cultural codes such as Anne Turner's yellow starched collar and the Devil's pricking, Fitzdottrel leads Justice Eitherside to draw the conclusion that Fitzdottrel is possessed by the soul-possessing Satan.

The courtroom scene echoes other scenes in the play where the projector spins a tale for an unsuspecting audience member who takes the spun tale for the truth. Quintilian explains how the orator can use words to alter the judge's reception of what is before him: "the appeal to the emotions will do more [than mere facts or proofs], for it will make them wish our case to be the better. And what they wish, they also believe" (6.2.5). Justice Eitherside, affected by the confusion regarding witchcraft and possession, wants to believe the drama being played for him is real, and is therefore unable to entertain contrary frames of reference that might shift the perspective through which he witnesses this drama. He takes the performance occurring before him at face-value and is unable to conceive of a fictional frame through which to view the words and actions in his courtroom. Unable to imagine fictitiously, Jonson's judge has no sense of humor. In light of Ramachandran's argument about humor promoting creative thinking, we define a sense of humor as an ability to entertain alternative potentialities and ironies beyond the established limits of a given context(s); in Coulson's language, this might be conceived of as a competency with frame shifting. This ability to unveil hidden meanings and imagine alternatives by reconceiving the contexts or interpretive frames bearing on a situation might have helped Justice Eitherside see through Fitzdottrel's possession performance.

The discharging of justice in the court depends on the judgment of the judge, a judgment, Jonson suggests, which is dangerously impaired by a lack of a sense of humor. A judge needs to be able to entertain, comfortably digest, and process keenly stories and events that appear radically awry from the everyday. In other words, engaging *secundum imaginationem* seems to be, for Jonson, requisite to a judicious analysis and judgment on all legal matters; therefore, Jonson posits the value of circumstance and the need to examine each case's idiosyncrasies as much as its commonalities with precedent cases. Jonson had ample opportunity to hear the opinions of lawyers and scholars on legal issues, as he developed close friendships with such professionals between the years before writing *Bartholomew Fair* and *The Devil is an Ass*. In fact, in 1613, he read the puppet court scene from *Bartholomew Fair* to his friends at the Inns of Court, and had occasion to discuss the Deuteronomeic code referenced by the puppets – a popular theme of antitheatricalists of the time – with his friend, legal scholar John Selden. Selden despised the over-reliance on simple Biblical exegesis that pervaded legal and scholarly thinking of the time. He sent Jonson his own literal translation of the Deuteronomy verse used to support a prohibition on transvestism: "A man's armour shall not be upon a woman, and a man shall not put on a woman's garment" (qtd. in Riggs 198).[17] The King James translation, however, suggests a deviance in the pleasure such cross-dressing brings: "The woman shall not weare that which pertaineth vnto a man, neither shall a man put on a womans's garment: for all that doe so, are abomination vnto the Lord thy God" (Deuteronomy 22:5). Selden's gloss on Deuteronomy provided Jonson with an example of how it is necessary to be able to read the layering of codes in order to judiciously interpret the law, especially as it is informed predominantly by the Bible, an often equivocal document itself. As implied by Selden, Jonson looks to the interpreter, rather than the speaker, for the responsibility of comprehension.

In the Prologue of *The Devil is an Ass*, Jonson situates his audience in the role of judge and jury:

> Would we could stand due north, or had no south, / If that offend – or were Muscovy glass, / That you might look our scenes through as they pass / We know not how to affect you. If you'll come / To see new plays, pray you afford us room, / And show this but the same face you have done / Your dear delight, 'The Devil of Edmonton.' / Or if, for want of room, it must miscarry, / 'Twill be but justice that your censure tarry / Till you give some. And when six times you ha' seen't, / If this play do not like, the devil is in't. (19–26)

His language reverberates with courtroom rhetoric; "miscarry," "justice," and "censure" all sound like the language used by a lawyer in an opening statement. He asks that the audience as jury show the play the same "face" it gave the popular play *The Devil of Edmonton*. This suggests, as we have argued Jonson's comedy also implies, that information can be taken in through different "faces" or frames. Jonson calls attention to the relationships between performer and audience by articulating how the scene might be read if the actors faced the opposite direction. The audience, as judge, is given responsibility for its perception of the play. Moreover, Jonson asks that any judgment of his play be assessed in light of the performance of the judge, and that if the audience insists on performing, they should abdicate the role of judge.

Jonson emphasizes the illusion of the theatrical event being performed for the current audience by having Fitzdottrel plan to see *The Devil is an Ass* in his new cloak that afternoon, which Fitzdottrel cannot possibly do because that day's performance is currently in progress as he speaks those lines. This metatheatrical paradox has a number of possible implications: it emphasizes the quality of the text, particularly in light of the censorship of the play following its first performance, by suggesting that it is such a good play that even its characters are eager to see it; and it recalls Jonson's plea to the audience in the Prologue to properly perform the role of audience. Barish argues that Jonson repeatedly criticized the judgment of his audiences: "they are bent on instant gratifications of a kind he has little wish to supply, and are, in the nature of things, prone to be swayed by opinion rather than reason" (139). According to Jonson, an audience so concerned with performing themselves on the stage that they squeeze out the players should see the play "six times" before passing judgment on it, or, better yet, they should read it.

In 1616, the year in which *The Devil is an Ass* was first produced, Jonson published his collected works, a move that reflected his desire to have his plays read. In the "Dedication to the Reader" that precedes *The New Inn*, a play published in 1631, Jonson praises the reader as he impugns the judgment of the play's first audience of 1629: "Howsoever, if thou canst but spell and join my sense, there is more hope of thee than of a hundred fastidious impertinents who were there present the first day, yet never made piece of their prospect the right way" (332). The play's title page reinforces this privileging of reader over listener: "Now, at last, set at liberty to the Readers, his Ma ^{ties} / Seruants, and Subjects, to be iudg'd" (Kidnie xxxvi). Jonson's critique of the audience as judge undermines the argument made by Barish and others that Jonson was an antitheatricalist; the performance of the poetry may involve artifice and disguise, but may still be honest. Joel Altman argues that, for Jonson, "fiction making is a productive form of play that enables him to draw out truths that would be otherwise unavailable" (188). In "An Epistle to Master Arthur Squib," Jonson supports Altman's claim: "Deceit is fruitfull. Men have Masques and Nets, / But these with wearing will themselves unfold: / They cannot last. / No lie grew ever old" (qtd. in Barish 144). The audience/judge, however, can either perceive these truths or not. Jonson asks his audience to be a good judge of a play that centers around the subjective reception of performance. Fitzdottrel fails to see the honesty of Pug's claim of being a devil and Justice Eitherside fails to perceive the dishonesty of Fitzdottrel's possession. For Jonson, however, there is always the opportunity of a *deus ex machina*, be it in the form of a character, or the intervention of a judicious reader.

The power of the performance must be matched by the ability of the judge to see through the performance, to read translucently. When convincing Fitzdottrel to fain possession for the judge, Merecraft explains that such a performance is common and simple: " 'Tis no hard thing t'outdo the Devil in: / A boy o' thirteen year old made him an ass / But t'other day" (5.5.49–51). Jonson most likely refers to a case of demonic possession in July of 1616 by thirteen year-old John Smith that caused nine women to be hung for witchcraft. What Jonson's audience would also remember is that James I questioned John Smith in August and found him to be faking possession and consequently released the remaining women Smith had accused.[18] Fooling a local judge, or audience, may be easy, but there is always the possibility of a king to effect justice at the eleventh hour. Mistress Fitzdottrel is also saved by a *deus ex machina*, though hers comes in the form of Satan. After Justice Eitherside delivers his judgment – "It is the Devil, by his several languages" (5.8.121) – the prison guard of Newgate arrives to report that Pug has escaped from prison, presumably to return to hell. Fitzdottrel then admits his deceit and, despite his claim that his wife is still a whore, Mistress Fitzdottrel is shown to be virtuous. When Jonson's audience is unable to view the play "six times" due to censorship, Jonson must hope for the *deus ex machina* of the reader of his works to repair the judgment delivered on his play by "fastidious impertinents" who judged his play the first time.

The Devil is an Ass merges the theater with the courtroom, so that the power in the theatrics of one troubles the supposed truth-value of the other. Fitzdottrel wants Merecraft's cloak to move him up the social ladder at the theater; for him, the theater contains the power to publish him as a finer gentleman. Fitzdottrel wants to go to the play in order to perform: "Come but to one act, and I did not care – / But to be seen to rise, and go away, / To vex the players, and to punish their poet – / Keep him in awe!" (3.5.42). Although Fitzdottrel never makes it to the theater as a member of the audience, he accomplishes his goal in the court. He uses the theatrics and story-telling of the theater in the courtroom to make himself the star of his own drama. Like Fitzdottrel in the courtroom scene, Jonson does not attempt to tell a story untouched by theatrical precedents, nor does he state explicitly how the audience – now in the position of Justice Eitherside – should interpret this story.

Just as the controversial case of John Smith haunts the courtroom scene, Jonson relies on the memories of other plays to inform the audience's reception of the play. Jonson's Prologue announces a number of plays he assumes his audience knows and to which meanings within *The Devil is an Ass* are indebted.[19] Jonson also refers to Shakespeare when Fitzdottrel refuses to be granted the title of Gloucester because of all the deaths he knows about from plays: "I ha't from the playbooks, / And think they are more authentic" (2.4.13). Throughout the play, as the editor Happé carefully traces in his notes, Jonson refers to, among other things, traditional morality plays, topical events, works of other playwrights (contemporary or deceased), and to his own *The Devil is an Ass*. He expects his play to be read in the context of the theater's precedents. The ghosts of *Macbeth* and *Dr. Faustus* in *The Devil is an Ass* haunt Jonson's argument regarding poetry and audience reception. Macbeth and Dr. Faustus, men felled by an inability to read through equivocation, provide guidance to an audience trying to decode Jonson's seeming attack on theater and the court of law.

Conjuring the equivocators: mediations with *Macbeth* and *Dr. Faustus*

We have seen that Jonson's play challenges social and legal codes through the use of comedic law in the courtroom. The deceit conceits employed throughout the play have shown us that characters can lie while apparently telling the truth – that is, equivocate – as a mechanism by which to propel others outside of their subjective territories. As we have seen, to comprehend an equivocation requires the same awareness of frame shifting and creative faculty as comedy. Understanding that Macduff, being "from his mother's womb / Untimely ripp'd" (5.8.15), is therefore not "of woman born" (5.8.13), means perceiving that "born" can mean more than one thing. It does not occur to Macbeth that the witches meant "born" in terms of a descent down the birth canal; he hears it only as referring to the pivotal role of women in producing men. In Coulson's terms, this dual understanding involves a similar frame shifting as a joke. Ramachandran might say the reanalysis of the original information in a new light produces a diminution of threat (though not for Macbeth) and a paradigm shift. Macbeth understands the "double sense" too late, of course, and the joke is on him. Had Doctor Faustus understood the simplicity of his reading of Mephostophilis's definition of hell, "Why this is hell, nor am I out of it" (*Doctor Faustus* 1.3.76),[20] he might have rethought the bargain. Macbeth, Doctor Faustus, and Fitzdottrel all fall victim to the deceit conceits of others. From Marlowe's successful projector Mephostophilis to Shakespeare's confusing weird sisters to Jonson's Merecraft, the plays endow words with the power of projective transversality. The plays cast doubt on, in Macbeth's words, "th' equivocation of the fiend / That lies like truth" (5.6.43). *Macbeth* and *Doctor Faustus* provide Jonson's play with spectral reminders of the role of a sense of humor in perceiving deceit conceits and reading through the often-confusing performance of projection.

The Devil is an Ass constitutes what transversal theory refers to as an "articulatory space," a discursive interface where symbolic discourses and social performances commingle and are imbued respectively and relationally with meaning.[21] In this bewitching space, *Doctor Faustus* is conjured by Jonson and participates in his discussion of performance, Satan, and equivocation. Before finalizing his agreement with Mere-craft, Fitzdottrel asks "let me repeat the contract briefly" (1.6.63); after reiterating the terms of their deal, he finishes: "This is your covenant?" (1.6.69). Jonson's switch from "contract" to "covenant" moves Fitz-dottrel's bargain from sounding like a legal business deal to evoking the covenants one made with the Devil. Popular enough to warrant a second printing in 1616, *Doctor Faustus* provides background informa-tion on the power of covenants and contracts to ensnare ambitious men. Jonson's audience, likely to have been familiar with the outcome of Doctor Faustus's contract, might then read Fitzdottrel's downfall as presaged in such a contract. While Fitzdottrel is not violently carried off to hell as Doctor Faustus is, Pug, Satan's minion in *The Devil is an Ass*, is. Pug is ineffective in sowing vice, crimes, or even doubt in London, and begs Satan to take him back to hell. Unlike Mephosto-philis, Pug is powerless. Jonson juxtaposes Pug's devilish incompetence with the impressive skill of Merecraft and Wittipol. These projectors are capable of convincing Fitzdottrel to sacrifice his money, wife, and social-standing for the illusions they promise him.

Mephostophilis uses less deceptive equivocation than Merecraft to convince Doctor Faustus to sign the contract in blood, relying instead on the promise of small deceit conceits "to delight thy mind" (2.1.85). Mephostophilis tells Doctor Faustus that if he goes back on his word to Lucifer, he will be torn to bits. What he does not mention is that Doctor Faustus will be torn to bits either way. Had Doctor Faustus been able to see through this equivocation, he might have followed the advice of the Good Angel and repented. As in *The Devil is an Ass*, in *Doctor Faustus*, the conjurer becomes the con-man. Granted twenty-four years of power for his soul, Doctor Faustus does not rid towns of plagues (1.1.21) or make man eternal (1.1.24) as he originally planned; he plays tricks on inferiors for the delight of his superiors. In the bor-rowed robes of the cardinals, Doctor Faustus and Mephostophilis tell the Pope that Bruno is to be imprisoned, not crowned. The Pope takes them at their word, and, believing their performance, becomes the duped Fitzdottrel, Justice Eitherside, or the "Simpler" of Dekker's "comedies." The Horse-courser asks Carter if he too had been conned by Doctor Faustus: "Did he conjure thee too?" (4.6.23). Before being struck dumb by Doctor Faustus, the Horse-courser insults him as "whoreson conjuring scab" (4.7.121). Fitzdottrel believes in the Devil because he believes that the laws against conjurers prove their exist- ence; but Jonson's fool fails to see that conjuring is illegal because it engenders a belief in what is not there. Conjuring the Devil, like the cons of a projector, invents the Devil in its invocation. This was not a concept unfamiliar to Elizabethans, for in his *The Art of Juggling,* Samuel Rid writes about a notorious cony-catcher good enough to be considered a conjurer: "This Cuthbert is esteemed of some and thought to be a witch of others; he is accounted a conjurer, but commonly called a wise man" (280).[22]

Jonson and Marlowe, as well as Rid, link conjuring to a con. Doctor Faustus's tricks, like the Devil's intervention in the sow's abortion or cream souring, sow doubt and confusion. Mephostophilis claims that the room for this doubt opened up when Doctor Faustus studied the scriptures: "when thou took'st the book / To view the scriptures, then I turned the leaves / And led thine eye" (5.2.102). On the other hand, Doctor Faustus's black book on necromancy forms a counter Bible, and yet is also bound by contracts and systems. Faustus seeks to dismantle its order and fails. His black book, with objective agency of its own, like the Bible, spreads dangerous knowledge and invites the Devil into a scriptural study. The fear of reading books within *Doctor Faustus* alludes to the debate within the Reformation as to whether or not the public should be allowed to read the Bible without mediation from their priest. At the play's end, the Chorus tells the audience to beware of too much curiosity: "Faustus is gone: regard his hellish fall, / Whose fiendful fortune may exhort the wise / Only to wonder at unlawful things, / Whose deepness doth entice such forward wits / To practice more than heavenly power permits" (Epilogue, 23). The *secundum imaginationem* practiced by these "forward wits" may project a danger- ous new alternate space, unacceptable to the state machinery or, for that matter, God. In Jonson's comedy the subversive performance of conjuring is not contained to the forward wits, but is rather something a fool or thirteen year-old boy can do. Read within the subjective territ- ory affected by the popularity of *Doctor Faustus, The Devil is an Ass* is not an indictment on the actions of the potential conjurers and projec- tors, but on the audience and judges who fail to see through the smoke and mirrors of performance to truths within the fictions.

Though weary of the kind of "borrowed robes" (1.3.109) for which Fitzdottrel whores his wife or in which Doctor Faustus cons the Pope, Macbeth also conjures evil on the equivocal promise of advancement.

Macbeth's inability to perceive that being a king but not begetting kings does not bode well for his lifespan proves to be a fatal error in judgment. Macbeth's porter, imagining himself as gate-keeper of hell, articulates the limitations of the equivocator: "Faith, here's an equivocator, that could swear in both the scales against the either scale; who committed treason enough for God's sake, yet could not equivocate to heaven" (2.3.8). Denied by God but believed by man, this equivocator sounds like Merecraft, capable of conjuring deceit conceits to fit the needs of Fitzdottrel, Guilthead, and Justice Eitherside. Only a powerful judge can spot "th' equivocation of the fiend / That lies like truth" (Jonson 5.6.43). In light of the deceit conceits Macbeth and Doctor Faustus fall for, Merecraft's and Wittipol's con of Fitzdottrel seems almost reasonable, as does the con of Justice Eitherside into believing that Mistress Fitzdottrel is a witch.

The role of women in conjuring, seducing, and equivocating in *The Devil is an Ass* must be read along with and through the invocation of Shakespeare's witches in *Macbeth*. When Pug first asks to go to Earth, Satan assumes his plans involve meddling in domestic jobs, such as: "Entering a sow to make her cast her farrow" (1.1.9). The bad mother cow trope is used by Shakespeare in his witches scene: "Pour in sow's blood, that hath eaten / Her nine farrow" (4.1.64). Jonson links the ingredients in Tailbush's make-up to the witches conjuring brew: "The crumbs o' bread, goat's milk, and whites of eggs, / Camphor, and lily roots, the fat of swans, / Marrow of veal, white pigeons, and pine kernels" (4.4.21). The ingredients required become more and more outrageous until Tailbush describes the skinning, boning, and pulverizing of a hen. This make-up potion is a "Scale of dragon, tooth of wolf" (*Macbeth* 4.1.22) away from conjuring Hecate. Through evoking Macbeth's witches in the context of a discussion on beauty, Jonson asks his audience to read the performance of feminine charms as conjured performances not to be taken at face-value. Similar to the diminution of threat experienced by the audience when it understands how Macduff was not of woman born, Jonson's conjuring women become laughably harmless in their desire to mask themselves when read through the codes of Macbeth's conjuring witches. Fitzdottrel's rhyming cant during the courtroom scene about his wife bewitching him also calls to mind Shakespeare's witches: "She comes with a needle, and thrusts it in, / She pulls out that, and she puts in a pin, / And now, and now, I do not know how, nor where, / But she pricks me here, and she pricks me there" (5.8.49). Happé notes that Satan was thought to torment the bewitched by pricking. The popular construction of Satan's propensity for pricking is also echoed by Macbeth's witches: "By the pricking of my thumbs, / Something wicked this way comes" (*Macbeth* 4.1.44). The plays are further linked by the allusion to the witches in *Macbeth* in Justice Eitherside's reference to Mistress Fitzdottrel: "A practice foul / For one so fair" (*The Devil is an Ass* 5.8.52). Justice Eitherside may expect Mistress Fitzdottrel to be the kind of "withered, and so wild" (*Macbeth* 1.3.40) bearded creature that Macbeth meets on "So foul and fair a day I have not seen" (1.3.38). While the witches of *Macbeth* spin the projector's tale of advancement with devilish effectiveness, in *The Devil is an Ass*, women are as incapable of witchcraft as they are of conjuring a younger face.

Of course, *Macbeth* and *Doctor Faustus* are tragedies and involve powerful forces of Satan and his agents. *The Devil is an Ass*, on the other hand, suggests that the greatest power of evil rests with Londoners. Jonson juxtaposes Pug's devilish incompetence with the impressive skill of Merecraft and Wittipol in fomenting evil in London. These projectors are capable of convincing Fitzdottrel to sacrifice his money, wife, and social standing for the illusions they promise him. The ghosts of *Macbeth* and *Doctor Faustus* in *The Devil is an Ass* make the cons in Jonson's play comedies by comparison.

While in Macbeth and Doctor Faustus equivocation and conjuring are the tools of the Devil, in *The Devil is an Ass* they are the tools of the joker. In 1616, the year *The Devil is an Ass* was performed, Shakespeare died and Marlowe's *Doctor Faustus* was reprinted. The case of John Smith's possession, with its shocking conclusion discovered by the King himself, might have reinforced the condemnation of witchcraft and conjuring within the subjective territories of Londoners. Jonson assumes this familiarity when he alludes to the boy of thirteen who made an ass of the Devil (5.6.49), just as he relies on his audience's understanding of the signs of possession that Fitzdottrel performs in the courtroom scene. Jonson's play takes these cultural codes as a point of departure in his indictment of the audience as incapable of being proper judges at the theater. His incrimination of the audience as judges, however, also applies to judges who he indicts as poor audiences. Jonson uses the troubled medium of theater, via comedic law, to show that the perceived dangers of performance spread outside the walls of the playhouse and into the halls of the court of law. His play, then, is a deceit conceit: spinning a tale that encourages thought *secundum imaginationem* in order to perceive meanings behind the poetry. The ghosts of *Macbeth* and *Doctor Faustus* mediate comprehension of cultural codes, perplexing and diffusing typical constructions of witchcraft, Satan, and the theater. Such encouraging of skepticism about witchcraft makes *The Devil is an Ass* not only heretical, but also, ironically, Satanic in that it breeds doubt about religious structures, which might have been why the play was censored, after which Jonson stopped writing plays for over ten years.

Because *The Devil is an Ass* was censored after its first performance, it did not receive the six performances Jonson felt his audience needed in order to deliver its judgment. Our fugitive reading, however, gives *The Devil is an Ass* the benefit of the doubt, revealing it to have been an ideal candidate for censorship, containing comedic devices that shake the dominant ideology of early modern England's state machinery and official culture. To be sure, in an unexpected alliance with the anti-theatricalists who eventually overthrew the monarchy, the same King James that accurately judged the performance of John Smith as bogus might have judged Jonson's play to be too dangerously questioning and bewildering of the Biblical law on which his own sovereignty resided.

Notes

1. For more on "objective agency," see especially *Becoming Criminal* (23–63), as well as *Performing Transversally* sic passim.
2. For more on "affective presence," see *Becoming Criminal* and *Performing Transversally* sic passim.
3. In *The Anatomie of Abuses*, Phillip Stubbes also refers to the theater as "Sathan's Synagogue" (143), and in *A Second and Third Blast of Retrait from Plaies and Theatres* (London: 1580), Anthony Munday calls "the Theater" "the chappel of Satan" (qtd. in Stubbes 302). For more on early modern English views of the public theater as an institution of the Devil, see Reynolds, "The Devil's House, 'or worse': Transversal Power and Anti-theatrical Discourse of Early Modern England."
4. For more examples and annotations, see editor Peter Happé's notes in Ben Jonson, *The Devil is an Ass* (Manchester and New York: Manchester University Press, 1994), 57–67. All quotations from *The Devil is an Ass* are from this edition.
5. For an especially savvy and persuasive reading of Thomas Dekker, John Ford, and William Rowley's *The Witch of Edmonton* as "one of the most radical dramatic challenges to the legal and cultural production of witches in the Renaissance," see Dennis Kezar, *Guilty Creatures: Renaissance Poetry and the Ethics of the Authorship* (Oxford and New York: Oxford University Press, 2001), 114–38.
6. For information about the witch craze in Renaissance England, see Bryan Reynolds, "Untimely Ripped: Mediating Witchcraft in Polanski and

Macbeth," in Reynolds, *Performing Transversally;* Alan Macfarlane, *Witchcraft in Tudor and Stuart England: a Regional and Comparative Study* (New York: Harper and Row, 1970); and *The Damned Art,* ed. Sidney Anglo (London: Routledge and Kegan Paul, 1977).

7. For detailed discussion of "subjunctivity," see Bryan Reynolds, "Transversal Performance: Shakespace, the September 11 Attacks, and the Critical Future," in *Performing Transversally:* 1–28.

8. For a historical examination of the implementation and ramifications for theology, philosophy, and science of *"secundum imaginationem,"* see Edward S. Casey, *The Fate of Place: a Philosophical History* (Berkeley: University of California Press, 1998), especially 106–16.

9. For an in-depth discussion of "projective transversality," see Donald Hedrick and Bryan Reynolds, "'A little touch of Harry in the night': Translucency and Projective Transversality in the Sexual and National Politics of *Henry V,"* in Reynolds, *Performing Transversally,* 171–88.

10. See Mikhail Bakhtin, *Rabelais and His World* (Cambridge: MIT Press, 1968), 1–20.

11. See Henri Bergson, *Laughter An Essay on the Meaning of the Comic.* Trans. Cloudesley Brereton and Fred Rothwell (New York: Macmillan, 1913), 71–93.

12. See Mary Douglas, *Purity and Danger: an Analysis of Concepts of Pollution and Taboo* (London: Routledge, 1978), 90–116

13. See Jonas Barish, *The Antitheatrical Prejudice* (Berkeley: University of California Press, 1981), 132–54.

14. Jonson also uses the name of Prospero for the conjurer in *Every Man In His Humor.*

15. Dekker capitalizes on what, as Reynolds explains elsewhere, was the socio-cultural, newsworthy, and entertainment values of criminal culture:

> criminal culture's chroniclers openly and repeatedly acknowledge and extol criminals for their theatrical and rhetorical skills. In describing criminal praxis, contemporary writers frequently situate their accounts of cons and other practices within an overarching discourse on the development and value of a criminal aesthetic.[15] Yet, while criminal praxis emerges in this discourse as artistically creative and worthy of recognition, it is also treated, in comparable terms, as a technical discipline that is both rational and logical, like that of an applied science. (*Becoming Criminal* 120)

16. For more on Anne Turner, see Ann Rosalind Jones and Peter Stallybrass, *Renaissance Clothing and the Materials of Memory* (Cambridge: Cambridge University Press, 2000), 62–7; Happé footnote, 64; and Margaret Jane Kidnie, *Ben Jonson: The Devil is an Ass and Other Plays* (Oxford: Oxford University Press, 2000):, footnote, 477.

17. For more on Jonson's biography, see David Riggs, *Ben Jonson: a Life* (Cambridge: Harvard University Press, 1989), and W. David Kay, *Ben Jonson: a Literary Life* (Basingstoke: Macmillan – now Palgrave Macmillan, 1995).

18. See Happé footnote, 206.

19. Specifically, *The Merry Devil of Edmonton* (1602) and Dekker's *If This Be Not a Good Play The Devil Is In It* (1611).

20. From Marlowe's *The Tragical History of Doctor Faustus* in *Christopher Marlowe: the Complete Works*. J. B. Steane, Ed. New York, 1969. This and all subsequent quotations from *Doctor Faustus* are from this edition.
21. For more on "articulatory spaces," see Reynolds, *Performing Transversally*, 1–28.
22. Samuel Rid, *The Art of Juggling*, ed. Arthur Kinney (Amherst: University of Massachusetts Press, 1990), 280. This is presumably a reference to the same Cuthbert Cony-catcher who authored *The Defense of Cony-Catching* (1592). In his textual note, Arthur Kinney suggests that Rid uses Cuthbert as a generic term for cony-catcher (311).

Works cited

Altman, Joel B. *The Tudor Play of the Mind: Rhetorical Inquiry and The Development of Elizabethan Drama*. Berkeley: University of California Press, 1978.
Anglo, Sidney, ed. *The Damned Art*. London: Routledge and Kegan Paul, 1977.
Aydelotte, Frank. "The Art of Conny-Catching", *Elizabethan Rogues and Vagabonds*. New York: Barnes & Noble, Inc., 1967, first edition 1913.
Barish, Jonas. *The Antitheatrical Prejudice*. Berkeley: University of California Press, 1981.
Bergson, Henri. *Laughter An Essay on the Meaning of the Comic*. Trans. Cloudesley Brereton and Fred Rothwell. New York: Macmillan, 1913.
Bond, Ronald B., ed. Certain Sermons or Homilies. London, *1547*.
— A Homily against Disobedience and Wilful Rebellion. London, *1570*.
— *A Critical Edition*. Toronto: University of Toronto Press, 1987.
Carlson, Marvin. *The Haunted Stage: The Theatre as Memory Machine*. Ann Arbor: University of Michigan Press, 2001.
Casey, Edward S. *The Fate of Place: a Philosophical History*. Berkeley: University of California Press, 1998.
Coulson, Seana. *Semantic Leaps: Frame-Shifting and Conceptual Blending in Meaning Construction*. Cambridge: MIT Press, 2001.
Dekker, Thomas. *The Guls Hornbook and The Belman of London: In Two Parts*. London, 1608.
Douglas, Mary. *Purity and Danger: an Analysis of Concepts of Pollution and Taboo*. London: Routledge, 1978.
Foucault, Michel. *The History of Sexuality: an Introduction, Volume 1*. Trans. Robert Hurley. New York: Vintage Books, 1990.
Gosson, Stephan. *The Schoole of Abuse*. (1579), rpt. London: The Shakespeare Society, 1853.
— *Playes Confuted in Five Actions*. London, 1582.
Greenblatt, Stephen. "Introduction", *Macbeth, The Norton Shakespeare*. Eds. Stephen Greenblatt et al. New York and London: Norton, 1997.
Greene, John. *A Refutation of the Apology for Actors*. London, 1615.
Harris, Jonathan Gil. *Foreign Bodies and the Body Politics: Discourses of Social Pathology in Early Modern England*. Cambridge: Cambridge University Press, 1998.

Harris, Max. *Theatre and Incarnation.* Basingstoke: Macmillan – now Palgrave Macmillan 1990.

Heywood, Thomas. *An Apology for Actors* (1612). Reprint, London: Shakespeare Society, 1853.

Jonson, Ben. *The Devil is an Ass.* Ed. Peter Happé. Manchester and New York: Manchester University Press, 1994.

Kezar, Dennis. *Guilty Creatures: Renaissance Poetry and the Ethics of the Authorship.* Oxford and New York: Oxford University Press, 2001.

Kidnie, Margaret Jane. "Introduction", *Ben Jonson: The Devil is an Ass and Other Plays.* Oxford: Oxford University Press, 2000.

Kopel, David. "The Intelligent Man's Guide to Lying Under Oath", *Inside Liberty* 8:3. Mar. 1999.

Macfarlane, Alan. *Witchcraft in Tudor and Stuart England: a Regional and Comparative Study.* New York: Harper and Row, 1970.

Marlowe, Christopher. *The Tragical History of Doctor Faustus* in *Christopher Marlowe: the Complete Plays.* Ed. J. B. Steane. New York: Penguin, 1969.

Quintilian. *The Institutio Oratoria.* Trans. H. E. Butler. Cambridge: Cambridge University Press, 1921.

Rainoldes, John. *Th' Overthrow of Stage-Playes.* London, 1600.

Ramachandran, V. S. *Phantoms in the Brain: Probing the Mysteries of the Human Mind.*
New York: Quill, 1998.

Reynolds, Bryan. *Becoming Criminal: Transversal Performance and Cultural Dissidence in Early Modern England.* Baltimore: Johns Hopkins University Press, 2002.

— "The Devil's House, 'or worse': Transversal Power and Antitheatrical Discourse in Early Modern England." *Theatre Journal* 49 (1997): 143–67.

— *Performing Transversally: Reimagining Shakespeare and the Critical Future.*
New York: Palgrave Macmillan, 2003.

Samuel Rid, *The Art of Juggling,* ed. Arthur Kinney. Amherst: University of Massachusetts Press, 1990.

Riggs, David. *Ben Jonson: a Life.* Cambridge: Harvard University Press, 1989.

Shakespeare, William. *Macbeth.* Ed. Kenneth Muir. London: Methuen, 1951.

Stubbes, Phillip. *The Anatomie of Abuses.* London, 1583.

Wills, Gary. *Witches and Jesuits: Shakespeare's Macbeth.* Oxford and New York: Oxford University Press, 1995.

5

I Might Like You Better If We Slept Together: the Historical Drift of Place in *The Changeling*

Donald Hedrick & Bryan Reynolds

Infinite space ... is space that expands endlessly, knows no term, has no limit, and finally engorges place in its massive maw. Even as dedivinized and thus as coextensive with the physical universe, the generality and openness of infinite space – in contrast with the enclosedness and particularity of finite place – have become virtually irresistible by the time we reach the threshold of the early modern era.

(Edward Casey, *The Fate of Place* 129)

DE FLORES: All this is nothing; you shall see anon
A place you little dream on.

(*The Changeling* 3.2.1–2)

Among the momentous historical sea-changes of the sixteenth and seventeenth centuries, which included the Copernican system, the great vowel shift, the public theater in England, nascent capitalism, and colonial expansion, there was an important conceptual revolution that has been less attended to. It is, moreover, one that connects all of these revolutions. We refer to the shift in the concept of place, particularly as it relates to the concept of space. While the transformation was multifaceted and far-reaching, one particular aspect of it has been charted for the history of ideas by Edward Casey as a shift toward a downplaying of the concept of place as it had been construed in Aristotelian/Ptolemaic and Christian philosophical traditions. Whereas "place" was, for Aristotle, largely thought of as the abstraction of containment (place is that which is contained), and a significant interpretive category in itself, the concept was enlarged with respect to Christian theology that viewed God as omnipresent, with place sometimes viewed as having power in itself, and sometimes even regarded as a sentient being.[1]

With the advent of the scientific revolution, nevertheless, place, although still granted its relation to the immeasurability of God, began to be subsumed within an idea of infinite space that was more scientific and mathematical, and that became synonymous with the infinitude of God. Place became a mere location on the expanse of a grid, and was therefore commensurately devalued as something without any particular power, and neutralized in effect by this historical move toward a scientific as opposed to a religious form of relativism. The later end of this trajectory of the concept, for Casey, is the current poststructuralist moment in theory, where locality and place seem to recover a version of their earlier importance and power, as we may see in the thinking of Gilles Deleuze and Félix Guattari on nomadism and on "immersions" in place.[2]

This important topic has not been treated at any length in scholarship on Elizabethan and Jacobean drama.[3] What is remarkable about Middleton and Rowley's play, *The Changeling*, and heretofore apparently unnoticed, is its embodiment of this historical shift, or rather its reaction to the decay of the local, and thus its reflection of the attendant anxieties, which worked into the already anxious and conventionally bleak moral universe of Jacobean tragedy, whose plots tend to make difficult if not impossible traditional "moral" readings of the plays.[4] *The Changeling*'s embodiment of the shift is so pronounced as to suggest that the play may constitute a conscious exploration of the fugitive concept of place-in-drift. The play pursues its own transversal investigation by which a place may be creatively altered – interpretively and experientially – in response to the openness and reduction of definition that constitutes new ideas of an overarching, subsuming, or all-encompassing space. An understanding of how "place" works in *The Changeling* will give us a deeper interpretation of the moral universe the play depicts.

Places of becomings, becomings of places

The term "place," indeed, occurs some thirty-five times in *The Changeling*, but this curiously large number alone does not give a full sense of the metaphorically perplexing and strange ways in which the term is used, nor of the theatrical or stage sense its usage emphasizes. Criticism of the play has tended to seize upon the less strange conceptual ramifications of the title through the term *changeling*, and its sense of *change* and *exchange*. The play itself, with its multiple "changelings" (Randall 349), sufficiently warrants these interpretations, as its moralizing and summarizing conclusion, given by Alsemero, reinforces the title: "Here's beauty changed / To ugly whoredom; here, servant obedience / To a master sin" (5.3.207–9). Following this statement are a litany of character confessions to change, testimonies to the identity becomings and comings-to-be spurred throughout the course of the play: Antonio admits, "I was changed, too, from a little ass as I was to a great fool as I am" (5.3.214–15); Franciscus states, "I was changed from a little wit to be stark mad" (5.3.217); and jealous Albius's promise to his wife – whose comic subplot represents the resistance to temptation that contrasts Beatrice-Joanna's fall – that he will "change now / Into a better husband" (5.3.224–5). Argued to be a darker version of *A Midsummer Night's Dream*, the play represents the madness and metamorphoses that can occur through experiences of love and sexuality (Malcolmson 338), similarly highlighting the transversality of consciousness and desire that can manifest in effect of passionate forays into the conventionally "unreasonable" or "irrational."

Moreover, critics who follow the playwrights' explicit announcement of this theme explore particularly the relation of change to character, especially with regard to Beatrice-Joanna, whose name points, at the very least, toward duality (Randall 340). Change is then extrapolated to considerations of gender itself, with women identified with the expectation of their changeableness, and in effect constituting the meaning of the term changeling (Burks 182) or further representing women's desire itself as a threat to male power (Garber 27). The play may, therefore, both challenge male authority and be sympathetic to its victims (Malcomson 337). Further along the theme of change, it has been argued that language itself becomes the object of change (Bueler 95), and that meaning is similarly its object, with the play striving to "keep meanings in their place," but ultimately, as a kind of linguistic *noir* tragedy, failing to do so (Dawson 110).

While the thematic idea of transformation is a critical commonplace, we can observe a more drastic drifting during the course of the play. The opening lines immediately underscore place as a philosophical and moral *issue* or, rather, question, namely, what it means to Alsemero that he met the already engaged woman in the temple where he now happens to stand:

> 'Twas in the temple where I first beheld her,
> And now again the same; what omen yet
> Follows from that?
>
> (1.1.1–3)

Dismissing the momentary possibility that this fact is an "omen" (1.1.2), he instead concludes that "the place is holy, so is my intent," thereby comparing his situation to "man's first creation, the place blest, / And is his right home back, if he achieve it" (1.1.8–9). In the allusion here, Alsemero unintentionally forecasts the play's replay of the Fall. Place provides a rationalization for swerving "back"; it legitimates if not dictates action. Later, in the final scene of the play, he recalls his confident reading of the place, and its delusion, where the murders and infidelities now mean that "the place itself e'er since / Has crying been for vengeance – the temple / Where blood and beauty first unlawfully / Fired their devotion and quenched the right one!" (5.3.76–9). In the opening scene, however, as he remembers back to his feeling that, " 'Twas in my fears at first" (5.3.76–9), his question about whether or not that temple was a "good omen" undermines its semiotic grounding. This uncertainty produces an instability or drifting that appears in the following dialogue of the scene, as the topic shifts to a new uncertainty regarding a question of place, namely, what the wind's significance is for Alsemero's impending sea voyage.

Like Jasperino, who urges that "the wind's fair with you" and that he is thus "like to have a swift and pleasant passage," Alsemero simply contradicts this optimistic reading: "Sure you're deceived, friend; 'tis contrary. / In my best judgment" (1.1.13–16). When Alsemero pointedly claims in defense that "the temple's vane" turned into his face, and that "I know 'tis against me," the pointed moral metaphor is given not in terms of subjectivity but in terms of place. This can be seen in how Jasperino contradicts Alsemero's interpretation by saying, "Against you? / Then you know not where you are" (1.1.22–3). Not knowing where you are is, of course, a pointed indicator of the moral drifting of the play, but one that begins to place doubt on the very moral bearings that might define what a "drift" would be. One must know where one is in order even to be able to recognize that one is straying.

The opening scene proceeds by deliberately linking the idea of place to the other chief concepts of the tragedy: time, sight, and the body. In their first private conversation in the scene, observed suspiciously by Jasperino, Beatrice-Joanna and Alsemero trade banter about their new love in terms of the judgment of their sight; Beatrice-Joanna warns against the rashness of such judgement, and Alsemero amusingly calls attention to the fact that the love was a matter of two related bodily sites: both of his eyes (his and Beatrice-Joanna's) agreed on the matter, "both houses then consenting" (1.1.80). In another "drift," here of time rather than direction, Beatrice-Joanna notes in an aside about having "missed time" in meeting Alsemero, now her true love, just after the wedding plans were underway. The villain, De Flores, is introduced in terms of sight or vision, but this introduction is ironically staged as the opposite of love at first sight in that Beatrice-Joanna finds she cannot stand looking at the fellow. She describes to Alsemero the case of having an idiosyncratic "infirmity" in finding something to be "a deadly poison" which to other tastes is "wholesome" (1.1.112–13). Alsemero philosophizes significantly on the matter, generalizing the situation to taste as a whole, namely that "there's scarce a thing but is both loved and loathed" (1.1.125). What is significant about De Flores's attentions to her in their earlier relationship is that they are related in terms of sight; that is, in his obsession with her he continually *places himself* in physical locations that will be visible to her, to "tempt her sight" (1.1.131). Sight is controlled by the *positioning* of what is to be seen as well as the seer, by changing location of both bodies and subjective territories. She confesses that De Flores is on good terms with her father, which explains the awkwardness of her loathing for him, to which Alsemero remarks that De Flores is therefore "out of his place, then, now" (1.1.136). Position or place, literal or subjective or social, determines not only power but also ethical agency, or its lack, and once again place takes a performative position within the logic of *The Changeling*.

One could well read the transformation of place in *The Changeling* as a movement from the opening scene's temple, through what Beatrice-Joanna terms the "labyrinth" in which she finds herself when she discovers that De Flores has blackmail potential over her for his killing of her fiancé, to the final revelations and tragic deaths of Beatrice-Joanna and De Flores. These "lovers" (the term resonates with the complications of its own meanings in the play) are regarded symbolically as a courting couple playing the game of "barley-brake," a game in which the loser's location or space is termed "hell." In this respect, one might at first consider the journey of the play through its spacetime as one moving from point A to point B. The conventional ethical equivalent of this decline is the descent into hell, a generally linear movement despite the intervening image of the "labyrinth." As De Flores anticipates the heroine's fall, he describes it as a geometric expansion of place, that would give him hope that he himself might "put in for one" if she begins to thus drift exponentially:

> For if a woman
> Fly from one point, from him she makes a husband,
> She spreads and mounts then like arithmetic –
> One, ten, a hundred, a thousand, ten thousand –
> Proves in time sutler to an army royal.
>
> (2.2.60–4)

Hence, De Flores reflects the seventeenth-century monumental shift to space as geometric and arithmetical, tied to the loss of self-identification which symptomizes transversality. With the radical sense of place as a site of agency or power, however, we note a difference between the standard conception of a clear division of heaven and hell, and a direct pathway running between them. In its two-part structure, Beatrice-Joanna's own ethical journey replicates this drifting and undoing of linearity itself. In the first part of the play, she strays from the right path by finding a murderer for her fiancé, although it takes her completely by surprise when De Flores makes the apparently reasonable case that she is hypocritical when she protests that he has destroyed her honor: he responds in astonishment that a woman "dipped in blood" can also speak of her honor or "modesty" (3.4.127). In any case, her moral movement at this stage is still graphable: that is, it marks a traditional moral movement of descent or decline, perhaps a fugitive escape but not a fugitive nomadism or wandering.

But in both its further movements and Beatrice-Joanna's next "decline," the play anticipates two other dark versions of artistic tradition linking place to moral decay and ambiguity: namely, the Gothic castle and the mean urban streets of *film noir*. These later places are sites where wanderings that are morally *exploratory* may occur, the sort of problematized and radical questioning that hearkens back to Dr. Faustus' inquiries (see Chapter 4). As she submits to De Flores's blackmail of her reputation, Beatrice-Joanna begins the part of her dramatic journey, and the audience's comparable journey, where point A and point B are no longer identifiable. Such a new space may be inferred from her transformed subjectivity. As De Flores describes her submission to his demand for repayment for performing the murder, she betrays a release of sexual energy, "panting" in his arms, and foreshadowing the next direction of her drifting descent:

> Thy peace is wrought forever in this yielding.
> 'Las, how the turtle pants! Thou'lt love anon
> What thou so fear'st and faint'st to venture on.
>
> (3.4.169–71)

Her emotions become an undefined economic "project," a risk-laden "venture" (another distinct motif of the play) of her own self outside even the territory that was defined as "criminal." While "honor among thieves" is not a wholly novel concept by any means, here the play will make the audience experience this undecidability through a perverse version of *Romeo and Juliet*, a version in which love is produced from hate-at-first-sight. In transversal terms, the transversal space of her criminality has been transformed into another subjective territory with its own set of controls (such as good versus evil), which in turn is made available for a further transversal movement into the space of "loving" and even "honoring" her criminal servant/lover.

Moral swerving, comings-to-be whore

What happens is perhaps more strange, complex, and counterintuitive than might be accounted for by a more conventional psychoanalytic approach that views loathing as simply an inversion of loving, an "unconscious" attraction (Daalder and Moore 507; Stockholder 145). By the end of the play, her space having expanded exponentially as De Flores predicted, Beatrice-Joanna is with apparent sincerity passionately commending his "service" to anyone who will listen. When he contrives to start the fire in Diaphanta's chamber, to destroy the witness and accomplice to Beatrice-Joanna's deception of her husband, she is so impressed as to exclaim, "I'm forced to love thee now, / 'Cause thou provid'st so carefully for my honor" (5.1.49–50). This way, she gives over her morally prophetic fear that a fire in one chamber "may endanger the whole house" (5.1.35). As De Flores frantically begins his sordid business, her complete reversal is clear from her newfound praise of the man she earlier loathed:

> Already! How rare is that man's speed!
> How heartily he serves me! His face loathes one,
> But look upon his care, who would not love him?
> The east is not more beauteous than his service.
> (5.1.71–4)

Thus, the Elizabethan moral cliché about "bed-swerving" has become swerving in a far more radical sense, as the ethical and emotional fields themselves become transformed, as if place itself could achieve agency, as a transversal subject experiencing something outside itself. His first misdeed, the killing she "hires," had actually only reinforced the moral poles of the existing moral field. In that instance, both the temple and the "labyrinth" in which she finds herself are still linear spaces, defined by their exits, even when the exit to the latter is difficult and arduous to locate. But as she experiences a true "love" and a "service" for the killer, after having already killed for a lover, her moral space is fully relativized, and she experiences an "immersion" in it rather than a journey through it. Her first descent is truly a descent, maintaining the traditional boundaries, position, and geography of morality; her second descent obliterates the first, removing the boundaries and leaving her in what De Flores will later call a "mist of conscience" (5.1.62), a momentary lapse or "paused consciousness" when/where one knows not which way is which (see Chapter 1). This is the description he uses, in fact, when he has the vision of Alonzo's ghost which for a brief moment at the end of the play reconfigures moral boundaries, despite the imagery of the "mist." The play's rich irony about place gives "direction" to Beatrice-Joanna as she helps carry out the next stage of the drifting, to kill the witness Diaphanta with the fire to be set in her chamber by De Flores: "Fear keeps my sould upon't. I cannot stray from't" (5.1.59).

The closet into which Alsemero shoves both De Flores and Beatrice-Joanna after their crimes and relationship have been uncovered is the most representative space of the moral progression: it is in effect a sealed-off space, a space of no exit, the theatrical evolution of the hell-mouth of earlier medieval drama. There, what Alsemero bitterly describes as a "black audience" will observe the wicked lovers' coupling, as he pushes De Flores into the closet where Beatrice-Joanna, no longer a sight or spectacle of infamy, can only be heard: "Peace, crying crocodile! Your sounds are heard" (5.3.121):

> Take your prey to you. Get you in to her, sir.
> I'll be your pander now. Rehearse again
> Your scene of lust, that you may be perfect
> When you shall come to act it to the black audience
> Where howls and gnashings shall be music to you.
> Clip your adult'ress freely; 'tis the pilot
> Will guide you to the *Mare Mortuum,*
> Where you shall sink to fathoms bottomless.
>   (5.3.122–9)

While Alsmero in these lines maintains the more traditional sense of space and place by which Beatrice-Joanna will become a "pilot" going from Point Heaven to Point Hell, his allusion to the "bottomless" Dead Sea suggests the newly transversal space of the two new lovers. The undecidability and literal invisibility of location here points toward the closet as a transversal territory where anything whatsoever may occur, and indeed does: what at first may appear (or be heard) as lovemaking turns out to have been De Flores stabbing her, his own "love" now translated into a sublime realm combining lust, jealousy, service, and release, in an endlessly looped snuff-porn video for a future "black audience." Alsemero describes it ironically to the as-yet-unknowing Vermandero as a "wonder" he will show him. But the term "wonder" itself begins to take on a new appropriateness in describing the affective dimension of the couple that has no explicit terms to describe it outside the traditional moral condemnations of the husband.

The term "whore," a general Elizabethan term of opprobrium for any serious moral failing, especially sexual failing, is itself subject to a drifting, a forgetting of what it is. Of course, it is actually De Flores who is technically the "whore" of the initial action, since it is Beatrice-Joanna who would pay him for his "service." Another slippage in the term occurs later in the play, as Beatrice-Joanna becomes suddenly jealous when the servant she has hired to sleep with her husband on their wedding night, in order to establish in the dark her own presumed virginity, begins to "dally" longer at her task than Beatrice-Joanna had planned on or can tolerate. As Beatrice-Joanna exclaims, "Oh, me, not yet? This whore forgets herself" (5.1.25), she instinctively applies the same term, "whore," that will later be applied to herself. At the play's conclusion, her husband bluntly says, "You are a whore" (5.3.32). De Flores, when all is exposed, declares that Beatrice-Joanna really only has "one thing" to confess, " 'Twas but one thing, and that she's a whore" (6.3.116). Of course, technically, she is playing the role rather of panderer rather than of whore in hiring Diaphanta. Significantly, she accordingly views the substitution of Diaphanta for herself in bed as a substitution of *place:* "The bride's place, / And with a thousand ducats!" (4.1.127–8). The moral swerving here, ironically one of *not* sleeping with her husband, not to protect her (nonexistent) virginity but rather her "honor" in concealing her crime, is once again the sort of drifting of moral boundaries distinguishing the nontraditional second part of the play from the traditional descent in the first part: the earlier descent as definably from point A to point B, the latter one non-chartable. To be sure, the minor character Diaphanta is similarly caught up in her own moral drift, as she first wishes only to help her mistress, it seems, avoid the supposed fears of the marriage bed. But we have seen this sort of drifting earlier. Applying the tragic protagonist's rule to the minor character, we might begin to conjecture about alternative outcomes of her decision: will she herself find "love" after sleeping with Alsemero? As "service" has before turned into "love," will a similar outcome occur in her case, the moral driftings become contagious and vertiginous in the persistent logic of transversality?

Transversal drifting, infinite space

The subplot of *The Changeling* is significantly more dependent upon the logic of place. In opposing the distinct places of court to an insane asylum, the play thematizes through location the relation between kinds of subjective territories and the ways by which they are transversalized. While the latter realm, the realm of comedy as opposed to tragedy, would at first seem to be the place of immorality or even of "drifting," it actually comes to signify a location with rules and governance, although the weaker rules and more tolerant governance befitting comedy. The anticipated "transgression" of the play's comic subplot – that the disguised madmen in the asylum will conquer the virtue of the jealous asylum owner's wife – would, even if it occurs, not undermine the confident moral territory upon which comedy itself conventionally stands. The moral bearing of comedy assumes that the audience will always be given the confidence that things will work out for the best. Significant, too, is that this rather well-defined space is itself internally articulated into two distinct "wards," about which comic observations and distinctions are continually drawn: a ward for fools and one for madmen. As Lollio describes the madhouse to Isabella,

> LOLLIO: When you have a taste of the madman, you shall (if you please), see Fools' College o' th' side; I seldom lock there; 'tis but shooting a bolt or two, and you are amongst 'em. (3.3.34–6)

Later, he adds to the picture of distinct locations, and therefore of the distinctions that transfer readily to the "fools" and "madmen" of the main plot of the play. He speaks of the governance of the insane as requiring two separate "shepherds":

> I would my master were come home. 'Tis too much for one shepherd to govern two of these flocks; nor can I believe that one churchman can instruct two benefices at once; there will be some incurable mad of the one side, and very fools on the other. (3.3.204–07)

In a sense, the distinction between the two is a transversal one, involving two sorts of "drifts." The madman's transversality involves a drifting outside himself in identifty, his subjective territory now fluid or shifting from presupposed categories, in this case gender itself. Thus, Franciscus, in disguise as a madmen, claims, "I was a man seven years ago," and "Now I'm a woman, all feminine" (3.3.69–71), whereas the transversality of the also disguised Antonio (*"Enter ... Antonio, like an idiot"* [1.2.81]) is in fool's banter, the meaninglessness of speech and action and uncontrollable laughter, even to the point of jesting with a transversal mathematics, where place is lost in tautology:

> LOLLIO: Tony, how man is five times six?
> ANTONIO: Five times six is six times five. (3.3.156–8)

Such repetition, potentially endless, signifies that dimensional understanding of the universe to which the play points and resists: "How can one be, exist on/at a point, and also be pointless, simultaneously everything?"

Indeed, the drift of place in *The Changeling* reflects the historical drift away from place and the attendant anxiety of placelessness. This anxiety reverberates with a vengeance as Beatrice-Joanna and De Flores are placed in the final enclosure, Alsemero's closet. The closet in a way echoes the figurative language that Alsmero used earlier in happily anticipating with Beatrice-Joanna their wedding to come, seeing marriage as itself an enclosure or place:

> My Joanna!
> Chaste as the breath of heaven, or morning's womb
> That brings the day forth: thus my love encloses thee. (4.2.149–51)

Alsemero's embrace or enclosure of her with these lines causes place to become an action, the action of enclosing, a theatrical image that contrasts with the drifts upon drifts into alterity, madness, or folly that constitute Beatrice-Joanna's, and her era's, journey from definition to undefinition. As De Flores embraces Beatrice-Joanna in the closet, whatever happens there in effect happens in no place, unseen to the audience, so that undefinition of infinite space ultimately prevails. The love murder, if it is love and if it is murder, occurs in the dark, but in the dark in the light of the stage, with the audience left in the dark. This recalls what Jasperino prophetically tells Alsemero in the opening scene of the play: "Then you know not where you are" (1.1.22).

Notes

1. See Edward Casey, *The Fate of Place* (Berkeley: University of California Press, 1999), in which he writes: "A powerful sense of something genuinely new was emerging, most dramatically in the uninhibited speculations of Cusa and Bruno, but insistently as well in the more cautious ruminations of Patrizi and the imaginative ideas of Campanella, who held that space is capable of feeling and sensing. Campanella also believed that space seeks to expand at every opportunity" (129).
2. On the work of Deleuze and Guattari in this respect, see Casey, *The Fate of Place*, 301–09.
3. See Andrew James Hartley, "Social Consciousness: Spaces for Characters in *The Spanish Tragedy*" (*Cahiers Elisabéthains: Late Medieval and Renaissance Studies*, 58 [Fall 2000], 1–14); and Robert Weimann, *Author's Pen and Actor's Voice: Playing and Writing in Shakespeare's Theatre* (Cambridge: Cambridge University Press, 2000), 188–9.
4. On the difficulty with "moral" readings of Jacobean drama, see Jonathan Dollimore, *Radical Tragedy: Religion, Ideology and Power in the Drama of Shakespeare and his Contemporaries* (Chicago: University of Chicago Press, 1984), 83–109.

Works cited

Bueler, Lois E. "The Rhetoric of Change in *The Changeling*." *English Literary Renaissance* 14.1 (Winter 1984): 95–113.

Burks, Deborah G. " 'I'll Want My Will Else': *The Changeling* and Women's Complicity with their Rapists." Ed. Stevie Simkin. *Revenge Tragedy*. Basingstoke: Palgrave Macmillan, 2001. 163–89.

Daalder, Joost and Anthony Telford Moore. " 'There's Scarce a Thing but is both Loved and Loathed': *The Changeling*." *English Studies* 80.6 (December 1999): 499–508.

Dawson, Anthony. "Giving the Finger: Puns and Transgression in *The Changeling*." *The Elizabethan Theatre* 7 (1993): 93–112.

Garber, Marjorie. "The Insincerity of Women." Ed. Valerie Finucci and Regina Schwartz. *Desire in the Renaissance: Psychoanalysis and Literature*. Princeton: Princeton University Press, 1994: 19–38.

Malcolmson, Cristina. " 'As Tame as the Ladies': Politics and Gender in *The Changeling*." *English Literary Renaissance* 20.2 (Spring 1990): 320–39.

Middleton, Thomas and William Rowley. *The Changeling*. *English Renaissance Drama: a Norton Anthology*. Eds. Katherine Eisaman Maus and David Bevington. New York: Norton, 2002. 1593–658.

Randall, Dale B. J. "Observations on the Theme of Chastity in *The Changeling*." *English Literary Renaissance* 14.3 (Autumn 1984): 347–66.

Stockholder, Kay. "The Aristocratic Woman as Scapegoat: Romantic Love and Class Antagonism in *The Spanish Tragedy*, *The Duchess of Malfi* and *The Changeling*." *The Elizabethan Theatre* 14 (1998): 127–51.

6

Fugitive Explorations in *Romeo and Juliet*: Searching for Transversality inside the Goldmine of R&Jspace

Bryan Reynolds & Janna Segal

> Every time I tell you I'm a real Romeo,
> It fills my soul and heart with anger, pain and sorrow.
>
> Sublime, "Romeo"[1]

The lyrics from the punk/ska/reggae band Sublime's song "Romeo" are indicative of popular conceptions of *Romeo and Juliet* as a tragic love story, and are suggestive of a number of meanings associated with the play's title characters that have permeated what some might dismiss as the lower depths of popular culture. In the song, lead singer and lyricist Bradley Nowell rejects his current partner who, although willing to please him sexually, fails to live up to his image of the perfect mate: "Want the kinda woman who can make me feel right / Not sloppy drunk sex on a Saturday night / Because I'm a Romeo – Romeo – with no place to go." Nowell implies that while being "a real Romeo" connotes manly charm, it also requires that one feels passionate, Petrarchan emotions, including "anger, pain and sorrow," over the absence of requited love. The singer/wanna-be Romeo's anguish is also a lament for what he imagines is his lack of self-cohesion. Obtaining "the kinda woman" that will secure his identity ("make [him] feel right") becomes imperative for "a real Romeo" to fashion a satisfactory self.

Although never mentioned by Nowell, Juliet, the female figure usually joined to the term "Romeo" with an "and,"[2] is invoked by the song's title and subject. Conjured in part by her absence, Juliet becomes the ideal "kinda woman" the singer – self-identifying as "a real Romeo" – longs for. Reciprocated love from the intangible Juliet is seen as the requisite ingredient for a process of self-authorship doomed from the onset, as evident by Nowell's lamentation, "Because I'm a Romeo –

Romeo – with no place to go." According to the lyrics, "a Romeo" cannot be *the* "real Romeo" without the symbolic and material presence of *the* Juliet, who is relegated to an always unattainable site for selfhood, a "place to go" to become the ideally "real" male identity that is, nonetheless, "no place" in reality. The patriarchal imperatives of the song are accompanied by a heteronormatizing framework explicit in Nowell's specific search for "a woman who can make [him] feel right." As suggested by the alliteration of "right" and "Romeo," without "a woman," wrong is what this/a/the Romeo will "feel."

The longing male subject and the desired female object of the song's narrative reflect normativity and formalism. The dichotomous tendencies of the song's narrative lead to the heart of the matter of "a Romeo's" story: this Romeo who longs to "feel right" is invested in dominant, idealized notions of romantic love that Shakespeare's lovers have come to embody. What "a Romeo" explicitly identifies as wrong – Dionysian "sloppy, drunk sex" – is a lament for his lack of the inverse, that Apollonian clean, sober love invoked by the fictitious figure the singer can only be "a" version of, and by the similarly illusive female attached to *the* "real Romeo."

With its debasement of sex as bacchanalian in contrast to *a* "real" love so pure it transcends the physical and reaches *the* romantic "right" believed to have been achieved by Shakespeare's Romeo and Juliet, Sublime's song testifies to the circulatory power of these iconic lovers who, apparently even within a music scene "alternative" to the mainstream music industry, can function as conductors for normative views. More important to the present analysis, however, is that the song is a strong exemplar of what transversal theory refers to as an "articulatory space," a discursive interface where symbolic discourses and social performances commingle and are imbued respectively and relationally with meaning. The specific articulatory space that the song exemplifies is what we call "R&Jspace," a conglomeration of the official and/or unofficial historical, political, cultural, and social spaces through which Romeo and Juliet resound in various manifestations, ranging from emblems of romantic love, legitimaters of forbidden desire, icons of teenage angst, and, in more recent critical incarnations, subversive agents of dominant ideologies substantiated by the names they themselves are so eager to doff.

Shakespeare's teenage lovers have come to emblematize formations of desire that have been officialized through spacetime in the image of their similarly iconicized author. The use of Shakespeare's play to reinforce a heteronormative conception of desire is an example of Shakespeare functioning as a "sociopolitical conductor," an ideologically-driven mechanism that works to create "subjective territory," which is an individual's interrelated realm of conceptual, emotional, and physiological experience. As Reynolds explains, the boundaries delineating subjective territory "are normally defined by the prevailing science, morality, and ideology. These boundaries bestow a spatiotemporal dimension, or common ground, on an aggregate of individuals or subjects, and work to ensure and monitor the cohesiveness of this social body" (Reynolds, "Devil's House" 146). To create and maintain subjective territory, a society's "state machinery," an assembly of "sociopolitical conductors," including educational, juridical, familial, matrimonial, and religious institutions, functions, consciously or unconsciously, to consolidate "state power" (any power working to create cogency, whether in social or material structures), manifest a concept of "the state" (in this case, the totalized state machinery), and produce a dominant, "official culture" (145).[3] As a result, "Official culture's sociopolitical conductors work to formulate and inculcate subjective territory with the appropriate culture-specific and identity-specific zones and localities, so that the subjectivity that substantiates the state machinery is shared, habitually experienced, and believed by each member of the populace to be natural and its very own" (147).

In contrast, and contrary to, state-orchestrated subjective territory is "transversal territory," a multi-dimensional space encompassing, among other known and unknown qualities, the nonsubjectified regions of all individuals' conceptual and emotional range. "Transversal movement," the conscious or subconscious breach and transcendence of one's subjective territory into a non-delineated alternative(s), threatens the stability of the state machinery and its regulating, official culture. Propelled by "transversal power," which is any physical or ideational force that inspires deviations from the established norms for an individual or group, transversal movements indicate the emergence and inhabitance of transversal territory. This experiential territory of learning, expansion, and metamorphosis through which people traverse when they violate the conceptual and/or emotional boundaries of their prescribed subjective addresses is a space people inhabit transitionally and temporally when subverting the hierarchicalizing and homogenizing assemblages of any governing organizational structure. Because it necessitates a departure from or subversive intersection with subjective territory, occupation of transversal territory opposes state power and threatens "official territory," the ruling ideology, propriety, and authority that provides the rationale, infrastructure, and parameters for a society.

Whereas Shakespeare's functions as a conductor for the construction and preservation of official territories are evident by his seemingly unshakable occupation of the Western canon,[4] his uses as a conductor for challenges to state power are equally recognizable in many of the diverse manifestations in which he has appeared and reappears. Responding to Shakespeare's past, present, and future workings as sociopolitical conductors of transversal power, Donald Hedrick and Bryan Reynolds have coined the term "Shakespace," which accounts for the articulatory space through which Shakespeare moves. Shakespace is useful for analyzing the various manifestations of Shakespeare since it indexes the elongated history of the official, transversal, and "subjunctive" – the imaginable "what ifs" and "as ifs" – of Shakespeare. As with other icons, such as Moll Cutpurse (see Chapter 2), whose work and/or "affective presence" – the combined material, symbolic, and imaginary existence of a concept/object/event – have spawned transversal spaces, the dissemination across spacetime of Shakespeare and his plays, propagandized as icons of official culture, has "encourage[d] alternative opportunities for thought, expression, and development," and has been used "to promote various organizational social structures that are discriminatory, hierarchical, or repressive" (Hedrick and Reynolds 9). For instance, *Romeo and Juliet* has functioned, as in the Sublime song, to popularize a notion of romantic love specific to heterosexuality; however, it has also, such as in queer adaptations of the text, worked to undermine that determination and stimulate transversal movements.[5]

Inspired by, often intersecting with, and able to compete against Shakespace,[6] R&Jspace accounts for the diverse social, cultural, political, and historical spaces generated, inhabited, and affected by Romeo and Juliet. In the aftermath of poststructuralism, the rhythm and tempo of travel in the Western university circuit of Shakespace has increased, generating in its path interconnected discursive traces and terrains, such as Hamletspace and Elizaspace; however, R&Jspace is one of the few Shakespearean-related articulatory spaces endowed with the reverberating power of Shakespeare. R&Jspace anticipates Shakespeare himself in the form of the play's primary source-text, Arthur Brooke's *The Tragicall Historye of Romeus and Juliet* (1562), as well as the alternative versions that succeeded Brooke's retelling of what Northrop Frye has deemed the archetypal love story (30);[7] and it continues to expand and appear in various media, including poetry, theater, novellas, song, cinema, television, and scholarship. Energized by recent trends in

Shakespeare studies, R&Jspace has, among scholars, become a battle-field on which critical perceptions and interpretative practices conflict, recalling the feuding-effect established in the first scene of Shake-speare's *Romeo and Juliet*.

To best articulate these conflicting discourses and envision a future-present path for potential workings through R&Jspace, we will engage in "fugitive explorations" of scholarly archives and Shakespeare's arch-ival text. Fugitive explorations, a concentration of transversal poetics' "investigative-expansive mode of analysis,"[8] call for readings of texts (of any subject matter) that seek to defy the authorities over them, authorities that can be found in all readings and reading environ-ments, both of a given text's or texts' inception and point of reception, their specific interpretive communities and those that constrain them, including the authorities by which a text's or texts' situational history is instilled with meanings. Hence, fugitive explorers endeavor wherever they are drawn, reparameterizing accordingly, to uncover fugitive ele-ments – narrative, thematic, semiotic, and so on – of the subject matter under investigation and the environments in which it has been con-textualized, particularly those that pressurize the authorities. Fugitive explorers, committed to empowering the elements on the run by giving agency where agency has been evacuated or forbidden, examine the elements and the trails they leave, endowing them with transversal or state power by spurring them on future attempts to avoid contain-ment. Giving representation to fugitive elements that may have been otherwise silenced, marginalized, or rendered invisible, fugitive explor-ations expose their covert activities, an operation that can sometimes work, inadvertently or willfully, to disempower or empower the once-hidden elements now locatable and thus vulnerable to deconstruction and/or co-option. Like all transversal thinkers, fugitive explorers look eagerly towards those future-present-spaces of potentiality where expe-rience is continuously immediate yet still foreseeable depending on available modes of observation and comprehension. It is with a hopeful glance towards a future-present that we pursue the golden lives, deaths, and resurrections of Shakespeare's "misadventur'd" lovers (1.P.7), who have transmigrated an "ancient grudge" and inspired a "new mutiny" in Shakespeare studies (1.P.3).[9]

Turbulence in R&Jspace

Insofar as *Romeo and Juliet* has been said to reflect and/or participate in clashes among "dominant, residual, and emergent" (in Raymond

Williams' terms) early modern ideological structures,[10] recent criticism of the play similarly engages and/or stages shifts in the "structures of feeling" (also Williams' term) of postmodern Shakespearean scholarship.[11] *Romeo and Juliet* has since the mid 1980s been a thoroughfare for materialist, psychoanalytic, and deconstructionist discourses on early modern England, all of which assert, in one form or another, that early moderners were aware of the constructedness of social identity and the whimsicality of self-formation, and thereby work to debunk the mythos of Renaissance individualism. Perhaps unsurprisingly for a text whose title characters revoke their interpellative names, defy the law of their feuding fathers in their very (instant) attraction to each other, risk eternal damnation (as threatened by the Church) by taking their own lives,[12] and whose lives are in turn rewritten into "the story of more woe" by the sole represented ruling monarch at the conclusion (5.3.308), the play has aroused interpretations from a range of post-structuralist positions, including Lacanian (Lehmann), Althusserian (Synder), Derridian (Belsey), and Marxist-feminist (Callaghan). There have also been powerful responses to these readings, responses that celebrate the lovers' transcendent and/or authentic love and resound with residual tones of pre-poststructuralist views of Shakespeare as the universal humanist whose work crosses spatiotemporal, social, cultural, and political divides (Bloom, Frye, Holland, Ryan, Wells). Emerging in the critical midst are fugitive explorations, such as the one that will be offered here, that track loose threads and follow the lovers to their textually-bound end and beyond in search of their prospects for the critical future, where they may break free of that "precious book of love [...]. / That in gold clasps locks in the golden story" (1.3.87–92).

Although our fugitive reading will differ significantly, it nonetheless supports a recent trend in criticism on *Romeo and Juliet* that, consciously or not, works to expose the ideological imperatives behind mainstream notions of romantic love by seeking to demystify desire as represented in the text. However inadvertently, critics from various camps (feminist, queer, new historicist, cultural materialist, psychoanalytic, and/or Marxist) have overtly or covertly reacted against the normativizing effects of the cultural imposition of Shakespeare's characters as icons of "true love" defined as heterosexual, monogamous, patriarchal, and/or worth dying for. Putting deconstructive pressure on the lovers' iconic status, for instance, Lloyd Davis argues that *Romeo and Juliet*, by engaging intertextually with conventional and unconventional early modern tropes of love, offers a hybrid, revisionist conception of desire "as the interplay between passion, selfhood and death"

(67). For Davis, this vision of desire as unable to transcend social mandates and as responsible for unfulfilled self-actualization disallows the moments of idealized romance in the play to be read as a ringing endorsement of romantic love.

Like Davis, Susan Snyder argues that desire in Shakespeare's play is intertwined and entrapped by the social, not disconnected from it. Contrary to her earlier reading of *Romeo and Juliet* as a tragedy of the "extraordinariness not so much in the two lovers as in the love itself, its intensity and integrity" ("Beyond Comedy" 132), Snyder subsequently contends that the play is a tragedy of ideology's inculcating effects/defects, with Romeo's and Juliet's deaths signifying the inevitable conclusion awaiting those who seek to depart from what in transversal terms would be described as their subjective territories: "For individuals who try to advance beyond their ideology but cannot undo its constitutive influence, there is no feasible way to live" ("Ideology and Feud" 95). For both Davis and Snyder, Romeo and Juliet's tragedy does not promote an idealization of desire as capable of overcoming conceptual boundaries; rather, it emphasizes the sociopolitical forces that inform and restrict the subjectivities of the desiring protagonists.

Alternatively, Jonathan Goldberg opposes interpretations of *Romeo and Juliet* whose heteronormative imperatives are implicit in their adherence to dichotomous differentiations of gender and sexuality that are, in his view, inapplicable to the early modern English period. Conducting what we would call a fugitive exploration, Goldberg argues that the play produces a "circuit of desire" (229) identifiable by tracing the series of "Rs" operating within the play. The open-ended "Rs," first initiated by the too often neglected Rosaline, produce a transgressive space of desire, or "sexual field" (228), which would have been identified as "the forbidden desire named sodomy" (228). Insisting that *Romeo and Juliet* "is a play about desire" (48), Catherine Belsey contends, like Goldberg, that it is anachronistic to read the play against traditional sexual-gender differentiations that did not emerge until the Enlightenment, but that it is not anachronistic to read the play as an articulation of Derrida's theory of supplementarity. To Belsey, the lovers' desire is the element that deconstructs the symbolic order of Verona: the lovers' bodies are emblazoned symbolically; they are only able to articulate their longings through the signifying systems of their culture, and yet the dominant order is unable to contain the deconstructive potency of their desire, which "seeks to overflow the limits imposed by the differential signifier" (54). Unlike Davis and Snyder, who maintain that *Romeo and Juliet* represents the containment of desire, for both Goldberg and Belsey, desire is the dissident element in the play that is disruptive to both early modern English social conventions and contemporary ideological structures vigorous in Shakespearean scholarship.

Antithetical to Davis', Snyder's, Goldberg's, and Belsey's explicit or implicit arguments for Shakespeare's and the play's subversive undermining of the voice(s) of authority co-opting the desiring teenagers in the play and/or in succeeding generations of performance, readership, and scholarship, Dympna Callaghan argues that the effect of Romeo and Juliet's love on the popular consciousness today is a reiteration of the text's initial "ideological function" (60), which was to naturalize an emergent formation of desire attendant upon the rise of capitalism in the early modern period, a manifestation whose "oppressive effects" include compulsory heterosexuality, homophobia, misogyny, and a false sense of autonomy (60). To apply transversal terms to Callaghan's reading, Shakespeare's play functions as a sociopolitical conductor of state power that manipulated and consolidated the subjective territories of early modern English playgoers and/or readers, and consequently shifted their social configurations against feudalism's imperatives in order to support the rise of capitalism and pervasiveness of royal absolutism.[13] Callaghan positions her denaturalizing project in opposition to a long history of scholarship on the period, the play, and the phenomenon of desire which has participated in ensuring that the "play's ideological project has become the dominant ideology of desire" (62). Her analysis also challenges essentialist constructions of Shakespeare as universal humanist and poststructuralist positionings of Shakespeare as subversive playwright by highlighting his participation, as authorial agent of *Romeo and Juliet*, in the cultural production of state-supporting ideological structures. Shakespeare's transhistorical value as articulator and authenticator of all things human is further called into question as the representation of desire in the play is linked to contemporary sociopolitics rather than "real" love. Nevertheless, Shakespeare's status as iconic Renaissance figure with transhistorical, official cultural influence is ultimately reaffirmed by Callaghan, whose interpretation of the play as "the inauguration of a particular form of sexual desire" that haunts us to this day (62) paradoxically redeems a text considered by some critics as a marred work, and, by extension, reasserts its author's transhistorical, inculcating power.[14]

Callaghan's reading becomes a focus of debate for Kiernan Ryan, whose analysis of the play's unveiling of language's inscribing power displays an awkward mixture of past and current approaches to

Shakespeare studies. Adopting a deconstructionist stance to reassert the value of an idealized form of love offered in the personages of Romeo and Juliet, Ryan proposes that the play's tragedy lies in the protagonists' inability to overcome ideological structures inscribed by language and sustain the "true identity" their love allows them to access (119). Ryan rejects traditional literary criticism (humanist, formalist, new critical), psychoanalytic approaches, and feminist interpretations of the text, which he regards as deterministic, essentialist, and/or reductive, and argues that the play's transgressive, agency-endowing quality lies in the lovers' creation of a reciprocal, free-from-subjugation form of love that, even though destroyed, affords the audience a view of an alternative expression of desire unburdened by the patriarchal power structures that delimit all other modes of desire represented in the play.

While Ryan dismisses Callaghan as a "crude radical critic" (126), his interpretation supports her assertion that contemporary scholarship has used the play to prescribe a heterosexualized conception of desire that renders alternative expressions abject. For Ryan, Romeo and Juliet's relationship arises as the preferable mode in the midst of other, malignant versions of desire, particularly Romeo's own Petrarchan "abject attitude" towards Rosaline, which is nothing more than "a degrading charade of domination and subjection" (119). Moreover, their union offers a vision of a futuristic, still unrealized formation of desire for the transhistorical, compulsively heterosexual couple: "The source of the play's abiding power lies in the way it foreshadows a more satisfying kind of love, freed from the coercion that *continues to drive men and women apart* and prevents their meeting each other's emotional needs" (118; our emphasis).[15] Ryan concludes that "the play vindicates the emergent right to love whoever one chooses, regardless of arbitrary prohibitions or prejudice" (124), yet his reading is riddled with statements on what constitutes perversion. Accordingly, for Ryan, the play condemns as perverse "a bond contaminated by the urge to use and to dominate, which perverts love into an instrument of pain" (124).[16] Shakespeare as author emerges from Ryan's analysis as the great humanist with a deconstructive edge, whose tragedies in general depict the destruction of innately "superior selves" by social forces (116), and who can articulate a predictable, futuristic variety of what Ryan seems to deem the ideal model of love: unfettered by patriarchal power-structures, yet compulsively heterosexual, and without the whips and chains of what Ryan dismisses as outrageous sexual perversion.

Ryan's interpretation reverberates with residual tones in its reaction against, and participation in, dominant "radical" criticism of the text;

nonetheless, unlike most reactions against poststructuralist approaches to the play's depiction of desire, his focus on the inculcating effects of language suggests an investment in recent critical approaches to the Shakespearean canon informed by the preoccupation with social constructedness that has become a trademark of postmodern theory. While Ryan's analysis can be said to commingle with various critical discourses, it adheres to the general strategy of current anti-postructuralist interpretations of the text, which has been to reimpress upon the reader the authenticity and/or sublime value of the title characters' mutual attraction. Such readings also challenge the de-authoring, de-centering tactics of postmodern approaches that jeopardize Shakespeare's iconic status by reasserting Shakespeare's universality as the articulator of the human emotion of love, itself envisioned as unrestrained by spatio-temporal, cultural, social, or political boundaries.

To give two examples of recent examinations of *Romeo and Juliet* that reassert the authenticity of the title characters' romance, we turn to Stanley Wells and Harold Bloom, who can be identified, respectively, as a moderate and a conservative in the Bardolatry party. Stanley Wells argues that *Romeo and Juliet*, the most "romantic" and the "bawdiest" of Shakespeare's plays (80), celebrates a formation of desire that is both ideal and real. Positioning the text in opposition to *Titus Andronicus*, which "was political in its implications," Wells finds *Romeo and Juliet* a "primarily private tragedy within a public setting" (76) that stages the reconciliatory power of true love, defined as capable of incorporating the physical while sublimely overcoming it. Wells juxtaposes Romeo and Juliet's romanticism with Mercutio and the Nurse's bawdiness, and argues that the protagonists' passion emerges triumphant because "it includes as well as transcends the physicality to which Mercutio and the Nurse are limited" (82). His formalist reading of the play fosters a normative ideology of desire in that it posits bodily expressions of desire as an inferior half of the equation of love by situating Romeo's and Juliet's emotions as superior to the "limited" "physicality" of their respective confidantes.

Arguing that Romeo and Juliet's "tragedy and their glory" is their "union" and its power to reconcile the feuding fathers (82), Wells implies that the text endows the heterosexual pairing with the power to resolve the differences between men, and thus works to consolidate a patriarchal order dysfunctional without the glorification of a heterosexual couple. Wells also dismisses alternative, less optimistic readings of the play's ending, such as the one that will be offered here, arguing that critics who deny "the qualified happiness" of the conclusion, albeit a "legitimate interpretation," distort the text by transforming "the play into social satire rather than tragedy" (83). As such, he denies any potential for subversion within the text by implying that its meanings are closed and that alternative readings of the play oppose the authority of the text, the author, and Wells as interpreter of the play's conclusive codings.

Wells' closed-book reading of the play is consistent with his contention that Shakespeare's works have an innate "abiding capacity to engage the minds and hearts of readers and theatregoers" (3). Interacting with the plays, readers and spectators are led to "both draw meanings from the texts and impose meanings upon them" (2). Led astray by the text itself, audiences can imagine meanings produced by, external to, and/or contrary to those of the scripted words of Shakespeare, whose authorship of such an engrossing body of work testifies to "the capacity of the human mind" (4). Shakespeare's "humanness" is here emphasized, yet his works continue to supersede audiences who, even when defying the authority of the texts, are only obeying its commands.

Wells' investment in Shakespeare as universal humanist is more moderate than Harold Bloom's, who, reclaiming Bardolatry as the pre-eminent secular religion, argues that the pervasiveness of Shakespeare is only comparable to that of the Bible, which shares "a certain universalism, global and multicultural" (3). An evangelical Shakespearean, Bloom contends that Shakespeare, that transcendental, universal "enigma" of "insoluble" proportions' (xvii), invented human personality, and that contemporary understandings of human consciousness stem from his works. Bloom dismisses dominant trends in Shakespeare studies as "French Shakespeare," and their practitioners as "professional resenters" and "gender-and-power freaks" who privilege theory over literature (9–10) and reduce Shakespeare to "a cultural phenomenon, produced by sociopolitical urgencies" (16). He endorses a return to a "more empirical mode" of criticism, albeit "staled by repetition," that insists upon what is "merely true": Shakespeare is "a more adequate representer of the universe of fact than anyone else, before him or since" (16).

Bloom dismisses recent criticism on *Romeo and Juliet* by the "commissars of gender and power, who thrash the patriarchy, including Shakespeare himself, for victimizing Juliet" (87),[17] and argues that the dramatic effect of this "tragedy of authentic romantic love" (97) is guaranteed by Juliet's "sublime state of being in love" (89). Echoing Wells, Bloom interprets Romeo and Juliet's "uncompromising mutual love" (89) as real and ideal, in that it is a "healthy and normative" desire (93) that transcends the loveless, strictly physical bawdiness of both the Nurse and Mercutio. Implicit in his reading of Mercutio as both an unknowing victim of an "authentic romantic love" that Mercutio disavows and as a repressed homosexual (97) is the presumption that real love is accessible only to the heterosexual couple, and will inevitably destroy the sexual nonconformist. Romeo's and Juliet's suicides are also lessons "in the catastrophes of sexuality" passed down from Shakespeare, and their demise makes apparent Shakespeare's recipe for the erotic, for he "invented the formula that the sexual becomes the erotic when crossed by the shadow of death" (89). The inevitable destruction by all forces – social, political, temporal, cosmological – of "a passion as absolute as Romeo's and Juliet's" emphasizes the erotic irony of the conclusion (102). Unlike Wells' celebration of the reconciliatory power of Romeo and Juliet's union, Bloom only finds "absurd pathos" in the final scene's enunciation of the still living characters' lives (103). Nonetheless, these "final ironies" do not undermine the play's romantic authenticity (103). For Bloom, *Romeo and Juliet* is "the largest and most persuasive celebration of romantic love in Western literature" (90), even if the "vision" it offers "of an uncompromising mutual love that perishes of its own idealism and intensity" depicts "mutual love" as idealistic, self-destructive, and vulnerable to all social and ephemeral forces (89).

Bloom's intention to revive a tradition of Shakespeare scholarship rejected by the followers of French theorists (and theorists of other national origins, we would add) is undermined by his own emphasis on Shakespeare's pervasiveness, which suggestively points toward official cultures' use of Shakespeare as a sociopolitical conductor. Bloom's assertion that, "Our education, in the English-speaking world, but in many other nations as well, has been Shakespearean" (13), suggests that the indebtedness of contemporary consciousness to Shakespeare's texts is due, at least in part, to Shakespeare's circulation in the educational apparatuses of state machinery. Like Wells, Bloom contends that audiences, subservient to the texts, cannot overpower them: "We need to exert ourselves and read Shakespeare as strenuously as we can, while knowing that his plays will read us more energetically still. They read us definitively" (xx). Readers and spectators, "definitively" inscribed by these reading texts, are enclosed within a subjective territory produced and reinforced by the circulation of Shakespearean conceptualizations through the machinations of state power. For Bloom, there can be no departure from these parameters, for we are made in the image of

Shakespeare's imaginings. Even those "resenters" who disinter socio-political energies in the texts are only "reflect[ing] the passions of *his* characters," namely Iago and Edmund (13). Apparently there is no means to escape from Shakespeare, who produced "an art so infinite that it *contains us*, and will go on enclosing those likely to come after us" (xix).

Disregarding Bloom's predilection for containment, fugitive explorers are not contained by Shakespeare's "infinite" art. Among such explorers is Goldberg, who chases after the elusive "Rs" to unravel what he regards as the "ideological function" of the "idealization of the lovers" at the conclusion of *Romeo and Juliet* (219). Like Goldberg, in the following pages we will chase after fugitive elements traceable in the text that undermine Romeo and Juliet's enshrinement within the play and in some critical approaches to it. Unlike Goldberg, our explorations recognize various manifestations – prescriptive and deconstructive – contributing to a collaboratively-authored articulatory space, and therefore our project does not function to reassert the notion of single authorship. Tracking the "Rs" through *Romeo and Juliet* to the *Sonnets*, and thereby delineating an intertextuality among Shakespeare's works, Goldberg productively argues that the play generates a field of desire potentially inhabitable by all genders. But Goldberg reaches expansively outside the text only so far as the Shakespeare canon. This self-imposed delimitation reemphasizes Shakespeare's individuality as single author, reasserts the image of Shakespeare as sole-authorial subversive agent, and stresses that any transgressive potential in Shakespeare's tragedy can only be uncovered by recourse to the *Complete Works of William Shakespeare*. We support Goldberg's persuasive reading and its relevance to Shakespeare studies, but we want to articulate R&Jspace within and without the confines of those texts attributed to Shakespeare. So far, we have pursued loose threads through various trajectories of R&Jspace, all along acknowledging its configuration as a collaboratively-authored space undergoing regular rewrites and always subject to specific interpretive communities. As we venture onwards into the text itself, we want to depart somewhat from the preoccupations of recent criticism on the play, and explore the less attended to metallic and monetary economy of "fair Verona, where we lay our [next] scene" (1.P.2).

Prologue to the "Golden Story" (1.3.92)

Shakespeare's lovers may see love as distinct from the nexus of business that nourishes Verona's state power, yet they too are constrained by the fluctuations of worth on an open market and the exploitative labor practices that have come to characterize capitalism. Going after the gold and other metals of monetary value circulating in *Romeo and Juliet*, we hope to make apparent the play's portrayal of love as commercial enterprise. In this regard we follow the demystifying trend in recent scholarship on representations of desire in the play, but with a critical difference. Whereas much postructuralist-informed criticism on *Romeo and Juliet* has focused on the characters' inculcation and victimization by sociopolitical conductors that define their desire, our aim is to expose the financial subterfuge we see operating in the text and show how it implicates the protagonists in the power dynamics of Verona's official culture, which, although pointing towards Romeo's and Juliet's incapacity to breach their subjective territories, also emphasizes their participation in maintaining the boundaries, real and/or imaginary, demarcating their lives. In our view, the co-opting of their union at the conclusion of the play by the figurehead of state power, Prince Escalus, is not necessarily as victimizing and/or glorifying as often perceived. Instead, the lovers become golden statues symbolizing the profit to be gained from the production of romance, as genre and ideology, while at the same time indirectly condemning that production as a corrosive metallic veil for the machinations of state power.[18]

Our reading of the play as a pronouncement of the commercial viability of romantic love critically expands on Hedrick and Reynolds' analysis of *Shakespeare in Love* (1998), the Academy award-winning film romanticizing Shakespeare's theatrical, literary, and amorous outpourings. Capitalizing on Shakespeare's "most beloved love story" and his "virtual identification with love itself" in the "current cultural sphere" (24), *Shakespeare in Love* creates biographical fiction by intercutting the theatrical and literary production of *Romeo and Juliet* with the development of both Shakespeare's career in the early modern English theater and his love affair with the oppressed aristocrat/inspiring actress/muse Viola De Lessups. The film articulates the heteronormative adage that "behind every great man there is a great woman" by imagining that without the love of Viola, Shakespeare would have remained the struggling, poetically-stumped hack-writer of *Romeo and Ethel, the Pirate's Daughter*, rather than become the genius who, as the film's portrayal of the opening production of *Romeo and Juliet* indicates, transformed the Elizabethan theater and its audience. Maintaining that the film departs from the Shakespeare play that concretizes its narrative, Hedrick and Reynolds argue that *Shakespeare in Love* portrays love as "a commercial sublime" instead of a transversal force (39). Instead of depicting business as functioning "in opposition to love (an opposition that both the poetry and the economic treatises of the early modern period might be evidence for)," or "romantically, as what must be transcended for love to occur," the film presents commercialism "as the source and model of affective relations," such as love (40). Unlike in *Romeo and Juliet*, where "the romantic concept of love that is Shakespeare's trademark [...] brilliantly overrides the play's sordid world and its jaded view" (23), *Shakespeare in Love* produces a late-capitalist articulation of love as sprung from commercialism, via the entertainment industry, where it continues to breed (40).

Hedrick and Reynolds fault *Shakespeare in Love* for its "failed transversal representation of love" (39) and its misrepresentation of the transversal effect of love's triumph in *Romeo and Juliet*. Alternatively, we find that the film reproduces the commercial effect in operation in Shakespeare's Verona. As in *Shakespeare in Love*, *Romeo and Juliet* represents business as the wellspring and model of love as it condemns the commercial as corruptive to the iconized lovers, who are unable to imagine or be imagined without reference to signifiers of commerce. Not functioning in opposition to love, the machinations of business can be seen to inform Romeo and Juliet's encounters and their eventual marriage, and ultimately to present their union as "golden." The lovers' attempts to move outside their subjective addresses, to think outside the commercial box, as it were, are thwarted in part by their own investments in the sociopolitical system, structured by mercantile exchange, within which they are confined. The play thus represents a failed attempt at transversality, as commercially-driven love is unable to conquer the official culture of Verona, which co-opts and commodifies its nonconformists. Simultaneously, the play capitalizes on the tragic effect it produces in its conclusion, which, in its attempts to secure for itself a long-lasting role in the early modern English entertainment industries (stage and print, primarily) and beyond, anticipates future manifestations of R&Jspace.

Merchants of Verona: hailings to the Prince

To show how the play works to enunciate the commercial viability of romantic love in Verona (and elsewhere) and critique the prescriptive practices of pre-capitalistic sociopolitical conductors, we must first consider in transversal terms the covert operations of the official culture in Verona in relation to its overt attempts to reaffirm itself in the wake of a rupture in its mechanizations caused by the Montague–Capulet feud. The portrait of Verona as a cohesive state is negated within the first scene of the play as servants from the opposing houses adopt the prejudices and defend, "The quarrel [that] is between our masters and us their men" (1.1.18–19). The servants' investments in their masters' "quarrel" testifies to what Snyder, in her Althusserian analysis of the play, defines as the familial feud's all-pervasive quality, whose every-whereness points to the feud's role as the ruling ideology in Verona: "Like ideology in Althusser's classic formation, the feud has no obvious genesis that can be discerned, no history. It pervades everything, not as a set of specific ideas but as repeated practices" ("Ideology and Feud" 88). Nevertheless, unlike in "Althusser's classic formation," where the family functions in conjunction with the dominant ideological state apparatus – be it the Church or the educational system[19] – the "ancient grudge" between both houses undermines the stability of the state (1.P.3), and thus can be identified as a transgressive ideology vying for dominance with the reigning ideology of the monarchy. In the language of transversal theory, the dueling families, rejecting their pre-scribed role as state-serving sociopolitical conductors, have become "resistive conductors" whose disruptive effects point towards the decentralization of Verona's state power. The transversal power of the feud is evident by its impact on "Verona's ancient citizens" (1.1.90), who have donned their weapons, "Canker'd with peace, to part your canker'd hate" (1.1.93). The families have transformed the state's elders from peaceful citizens to inept peacekeepers of "the quiet of our streets" (1.1.91), thereby reshifting social roles in their resistance to the "quiet" policies of Verona's official territory.

The feud's disruptive power is so threatening to the state that it pro-vokes a death-promising condemnation from Prince Escalus (1.1.79): "If ever you disturb our streets again / Your lives shall pay the forfeit of the peace" (1.1.94–5). The Prince's assertion that the families will "pay" for any future outbreaks of rebellion transfers the cost of main-taining civic order from the state to the families, whose lives will serve as recompense for the price of upholding peace. Monetary metaphors resound again when the Prince, following Mercutio's death, exiles Romeo: "I have an interest in your hearts' proceeding; / My blood for your rude brawls doth lie a-bleeding. / But I'll amerce you with so strong a fine / That you shall all repent the loss of mine / I will be deaf to pleading and excuses; / Nor tears nor prayers shall purchase out abuses" (3.1.190–5). The death of his kinsman Mercutio is expressed as a business venture securing Prince Escalus the profit of Romeo's exile:

it is an "interest" leading to a personal "loss" for which he will be compensated with Romeo's banishment, itself "a fine" that cannot be "purchase[d] out" by remorse ("tears") or recourse to God ("prayers"). Anticipating Marx's analysis of the dehumanizing effects of capitalist modes of production, in each of the Prince's sentences human lives are abstracted into compensatory commodities to cover the costs of disorder incurred by the state. The Prince's biocapitalistic tactics, whereby human bodies are rendered into consumable things whose consumption feeds the ruling order by beefing-up the state's official territory, imply that the production of state power relies upon a pool of subjects who will be commodified if and/or when they are disobedient. His royal disciplines and punishments reveal the workings of the play's industry-driven monarchy to ensure that Verona remains a well-run machine.

Whereas Prince Escalus' signifying practices reveal the state's interest in capitalistic enterprise, the monarch can also be seen as making an appeal to the more merchant-like than feudalistic fathers, particularly Capulet, whose propensity for profit-making and tyranny over his daughter foreshadows the comic cruelty of the later merchant-patriarch Shylock. Described by his illiterate servant as "the / great rich Capulet" (1.2.80–1), while himself down-playing his estate as a "poor house" to the suitor Paris (1.2.24), Capulet's monetary value fluctuates in the open market of Verona. His greed is somewhat confined by the laws dictating social propriety in Verona, as is evident when he prohibits Tybalt from dueling with Romeo at the Capulet feast, for since all of "Verona brags of him / To be a virtuous and well-govern'd youth. / I would not for the wealth of all this town / Here in my house do him disparagement" (1.5.66–9). Capulet's insistence that he "would not for the wealth of all this town" offend Romeo suggests the affluence of Verona, his own monetary preoccupations and aspirations to acquire greater financial status, and that he is less threatened by the Prince's ban on brawls than by social ostracization resulting from rudeness to a well-liked member of the Montague clan. Capulet's concern with financial acquisition and social acceptance can also be seen in his insistence that Juliet marry the Prince's kinsman Paris, that "gentleman of noble parentage" (3.5.179) who will ennoble Juliet, "Unworthy as she is" (3.5.144). Juliet, described by Capulet as "the hopeful lady of my earth" (1.2.15), is valued by her father for her potential as a vessel bearing his "earth" who, married into the monarchical order, can breed future profits that will in turn secure Capulet's sought-after, heightened social prestige in Verona.

Capulet's compulsive concern with investiture is repeatedly emphasized in the play, and the absence of similar moments involving Romeo's father implies that both houses are not "alike in dignity" (1.P.1), but rather that Montague enjoys a more stable financial security. The conclusion to the play, in which the feuding fathers bid for rivalry for the Prince's favor by projecting their now-deceased children into statues of "pure gold" (5.3.288), questions this seeming imbalance of "dignity" by illuminating the shared mercantile propensities of Capulet and Montague. Likewise, the conclusion functions to solidify the relation between commerce and state power in Verona implied by the Prince's previous employment of monetary metaphors.

Mirroring the biocapitalistic machinations of the Prince, the fathers abstract their respective children into signifiers of Verona's prosperity, golden figures whose "rate" will "be set" for the length of time that "Verona by that name is known" (5.3.299). The planned commodification of Romeo and Juliet into displays of official culture is a response to the Prince's reprimand of society for allowing the families' feud to incur a wrath that has gone so far as to impeach upon his private property, taking the lives of his "kinsmen" (5.3.294). Following the Prince's proclamation, "All are punish'd" (5.3.294), Capulet offers his "hand" to "brother Montague" (5.3.295), a peace offering immediately outdone by his former foe: "But I can give thee more, / For I will raise her statue in pure gold, / That whiles Verona by that name is known, / There shall no figure at such rate be set / As that of true and faithful Juliet" (5.3.297–301). Not to be overbid, Capulet responds by offering to construct an equally valuable version of Romeo: "As rich shall Romeo's by his lady lie, / Poor sacrifices of our enmity" (5.3.302–03). Romeo and Juliet have thus become objects in an auction of loyalty to the Prince, with each of their respective fathers using their children's images to publicly prove themselves more apologetic for the discord their feud has cost the state.[20] Moreover, the transmutation of the lovers into golden financial "figures" compensating for the loss accrued by the state simultaneously declares healthy, wealthy, and wise the socioeconomic structures supporting Verona's official culture. This deems a mound of gold far more valuable than a pound of flesh.

Verona then is not, as Snyder argues, "constituted by the feud, [which] asserts itself like any ideology as the only reality there is" ("Ideology and Feud" 93). Instead, it is constituted by feuding ideologies that generate dueling subjective territories whose common parlance is the language of commerce. Romeo and Juliet are caught between rivaling ideological structures that nonetheless share an investment in profiteering. They seek to transcend one set of boundaries through the auspices of the other (money, their servants, the Church), only to find themselves trapped within the machinations of the two realms they traverse between.

Das capital "L" is for "love"

Shakespeare's "pair of star-cross'd lovers" (1.P.6) are willing to die for their love, for each other, and in defiance of their subjective addresses as son of Montague and daughter of Capulet. Yet, as the First Act Prologue promises, their union has the potential to forge an alliance between "their parents' strife" (1.P.8), creating a bond, endorsed by Prince Escalus, between Verona's two dignified houses. As the lovers strive to contravene the subjective territory engineered by their families, they inadvertently work to solidify the parameters of Verona's dominant sociopolitical landscape. Calling upon Verona's less dignified subjects, namely the Friar and their respective servants, as well as Mantua's "beggarly" Apothecary (5.1.45), Romeo and Juliet cling to their Veronese social class identities, participate in the privileges of wealth afforded them and legitimated by the page-turning "precious book of love" circulating among Verona's literate class (1.3.87), and they live and die by the tune to which the heart of the city beats: money can buy you love in Verona.

Romeo and Juliet breach the prescriptions of their families reinforced through the ideology of the feud when they meet and fall in love, or, as the Second Act Prologue suggests, when they become "Alike bewitched by the charm of looks" (2.P.6). The Prologue presents love as a magical, captivating process inspired by surface appearances that lead to emotional transference and conceptual transformations, a process we would identify as transversal movement. This framing of love as bewitchment is supported by the mutative effects of love on the characters. After Romeo espies Juliet, Rosaline, Capulet's "fair niece" (1.2.70) and the former "precious treasure" of Romeo's affection (1.1.231), "is now not fair" to Romeo (2.P.4), whose desire has been transferred to a new object closer in relation to "rich Capulet" (1.2.81). Like the "uncharm'd" Rosaline (1.1.209), whose refusal to return Romeo's affections engenders a rupture in his subjective territory that he bemoans ("Tut, I have lost myself, I am not here. / This is not Romeo, he's some other where"

[1.1.195–6]), Juliet is endowed with transformative power over Romeo: "Now Romeo is belov'd and loves again" (2.P.5). No longer the mournful Petrarchan lover, Romeo becomes the "belov'd" object of another's affection, one who has the power, through the interpellative act of hailing, to alter Romeo's self-conception: "Call me but love, and I'll be new baptis'd: / Henceforth I never will be Romeo" (2.2.50–1). Capable of loving again, Romeo's bewitchment and its transformative effects suggest that love is a social, supplemental process of transference – not an isolated, privatized experience – whose "positive" outcome of reciprocation relies on the object's likewise enchantment. This transference similarly encodes Juliet's affection for Romeo when one considers that prior to their meeting, Juliet vows to peruse Paris, that "precious book of love" (1.3.87), at the Capulet feast: "I'll look to like, if looking liking move" (1.3.97). Looking, capable of inspiring "liking," engenders movement, and, in the case of Juliet's viewing of Romeo, it motivates a "move" away from her prescribed address as bride-to-be of the Prince's kinsman.

Mutually bewitched by appearances in the social act of looking, Romeo and Juliet recall the rhetoric used by early modern English antitheatricalists to describe the dangerously transformative power of public performance. As Reynolds explains in *Becoming Criminal*, the antitheatricalists recognized the infectious, transformative power of the public theater, where audiences, exposed to alternative conceptualizations of self and others through the act of looking, could themselves undergo the transversal process of becoming-other-social-identities.[21] Like audiences in the early modern English public theater, Romeo and Juliet, in looking and thereby becoming-other by example, make themselves vulnerable to subjective awareness outside their socially-prescribed selves. Both Romeo and Juliet attend the Capulet feast with the intention to look on love – Romeo to "rejoice in splendour" at the "sight" of Rosaline (1.2.102–03), Juliet to "endart mine eye" (1.3.96) on "Paris' love" (1.3.94) – and, following the logic of the antitheatricalists, each of them is predisposed to finding love in a configuration fugitive to that which is dictated by the state-supportive sociopolitical conductors guiding their perceptions. The formation of desire they find in the example of each other is characterized by its transversality: it is a "passion" that "lends them power" (2.P.13) to venture outside their subjective territories and become other than what they were, changing from, in the case of Romeo, a rejected, lovelorn teen to an exiled husband, and in the case of Juliet, an obedient daughter to a rebellious, teenage-wife. Regardless of the degree to which they move transversally and/or fugitively, the lovers remain emotionally, conceptually, and/or physically constrained by the "means" necessary to facilitate a legitimate union in the official territory of Verona (2.P.11; 13). Juliet in particular is restricted by the fact that her "means" in Veronese society are "much less" than Romeo's (2.P.11). According to the Prologue, "time" provides the protagonists with the necessary "means, to meet" (2.P.13); however, "time" is money in fair Verona where, through their "means" to their ends, Romeo and Juliet's passion is marshaled through Verona's state machinery and/or imbued by the ideologies it generates.

Romeo, who enjoys the privileges afforded to men of his class status in Verona, primarily conceives of love in business terms and seeks to secure the object of his affection, whether Rosaline or Juliet, through monetary means. Romeo's recurrent recourse to financial terminology reflects Verona's official culture, as well as his own contribution to concretizing "Verona walls" without which "There is no world" he can imagine "But purgatory, torture, hell itself" (3.3.17–18). Unable to conquer Rosaline with a "siege of loving terms" or "th' encounter of assailing eyes" (1.1.210–11), Romeo attempts to seduce her with currency, only to fail, for Rosaline will not "ope her lap to saint-seducing gold" (1.1.212). Although enticing even to a "saint," gold cannot buy love from Rosaline, whose refusal to be purchased only increases the value of her appearance: "O she is rich in beauty, only poor / That when she dies, with beauty dies her store" (1.1.213–14). Upon seeing Juliet at the Capulet's ball, Romeo again is entranced by the wealth invoked by the appearance of beauty, perceiving Juliet, "As a rich jewel in an Ethiop's ear – / Beauty too rich for use, for earth too dear" (1.5.44–5). Juliet's beauty is not "too rich for use," as Romeo learns from the Nurse, for it is through her use as a bride that the wealth Juliet signifies can be possessed: "he that can lay hold of her / Shall have the chinks" (1.5.115–16). Juliet later reaffirms the Nurse's understanding of marriage as a means to inheritance when she vows that, if Romeo marries her, "all my fortunes at thy foot I'll lay" (2.2.147). That Romeo's interest in marrying Juliet is at least partly motivated by profit is again suggested when Romeo solicits Friar Laurence to sanctify his marriage to "the fair daughter of rich Capulet" (2.3.54). Romeo recognizes that in Verona's official culture Church blessings are required to legitimate his purchase through marriage, and therefore convinces the Friar to assist him in securing his possession with the "holy words" (2.6.6) necessary for Romeo to "call her mine" (2.6.8).

Romeo's repeated proffering of money to secure his gains through the networks of an intact labor system similarly demonstrates his investment in the commerce-driven official culture of Verona. Sending the Nurse to tell Juliet of their impending marriage, Romeo offers the Nurse money "for thy pains" (2.4.179), with "pains" presumably referring to the physical and/or emotional ramifications of her labor, as well as to the emotional effects ensuing from the insults hurled on her by Mercutio, the Prince's kinsman. When the Nurse refuses even "a penny" from Romeo's purse (2.4.180), Romeo insists, impressing upon her his position of authority: "Go to, I say you shall" (2.4.181). Promising future financial rewards for her dedication to his cause – "Farewell, be trusty, and I'll quit thy pains; / Farewell. Commend me to my mistress" (2.4.188–9) – Romeo again reminds the Nurse of the power his financial wealth affords to "quit thy pains." Romeo purchases the Nurse's employment in a cause counter to the commands of her employer, Capulet, and he impels her to engage in a transgressive act by recalling the imperatives of an employer: be faithful to me, work hard for me, and you shall be financially rewarded by me. As the distributor of currency envisioned as having the medicinal properties to cease "pains," Romeo frames himself as the boss as well as a medicine man capable of providing the means necessary to alleviate the physical and emotional woes ("pains") of the lower-classes in Verona's social hierarchy.

In a similar scenario, Romeo, breaking into the Capulet tomb under the auspices of retrieving a "precious ring" from Juliet's "dead finger" that he "must use / In dear employment" (5.3.30–2),[22] buys his servant Balthazar's obedience, and by extension his participation in the criminal act of tomb-raiding. Romeo first reminds Balthazar of his authoritative position by threatening to "tear thee joint by joint" if Balthazar lingers to discover Romeo's intents (5.3.35), and then he tries to secure his servant's loyalty by proffering money: "Take thou that. / Live, and be prosperous, and farewell, good fellow" (5.3.41–2).[23] Like the Nurse, Balthazar has been drafted into active engagement in a conspiratorial act – this time against the Capulet clan as well as the Prince, who has exiled Romeo from Verona – through the provision of monetary compensation that works to end "pains," in this case alliteratively articulated as an endowment providing a "prosperous" existence. Having taken "that" from Romeo, Balthazar can now not only "Live," but he can "be prosperous." Romeo again emerges as distributor/employer, beneficiary/boss as he funds his transgressive projects with the bought labor of his and/or Capulet's minions.

In Mantua, Romeo no longer frames himself as a distributor of medicinally-charged currency with the power to purchase labor, as he does in Verona. Rather, he adopts a less altruistic attitude towards money, particularly gold, which he there deems abject. Currency generally retains its positivist qualities for Romeo when he is secure within his "Verona walls" (3.3.17). Although "saint-seducing gold" cannot buy Romeo entrance to Rosaline's "lap" (1.1.212), it is capable, with the promise of alleviating "pains" and affording prosperity, of securing the labor necessary to create a gateway to Juliet. Exiled from Verona, and thereby forced to live outside of its conceptual, emotional, and physical boundaries, Romeo reinscribes money as a corruptive "poison to men's souls" (5.2.80). This depreciation of money's medicinal powers is related to Romeo's displacement and instatement in a world ulterior to Verona. For Romeo, anywhere outside Verona is only conceivable by reference to the realms of punishment promised by the Church for sinning: "There is no world without Verona's walls / But purgatory, torture, hell itself" (3.3.17–18). Mantua, then, is that "no world" like home Romeo fears, a universe lacking Verona's heavenly riches where exiles, such as himself, are confined.

The exiled Romeo maps Mantua's sociopolitical body according to the dictates of his Veronese social class consciousness, and his distributory tendencies follow the behavioral patterns he displays in Verona. He still remains a man of Veronese means in the hellish, financially-depleted confines of Mantua. Clearly he has money to burn, for he offers "forty ducats" to the Apothecary in exchange for "A dram of poison" (5.2.58–9). Searching for "means" to reunite with Juliet in death (5.1.35), he remembers the Mantuan Apothecary as a man with "Meagre" "looks" (5.1.40) and "penury" (5.1.49). Romeo reads the Apothecary through the visual signifiers of poverty and deduces that "this same needy man must sell it me" (5.1.54). Romeo recalls his Veronese social class role by framing himself as the financially-endowed figure who can satiate the "needy man['s]" needs. Displaced, Romeo tries to carve himself a position of authority in Mantua by hailing the Apothecary according to his interpretation of the visual demarcations of social class: "I see that thou art poor" (5.1.58). The interpellative gesture empowers the exiled teen by impressing upon the Apothecary his subject position in relation to Romeo.

Veronese money and class status are apparently not enough to secure for Romeo an authoritative position in Mantua, as seen by Romeo's recourse to more Machiavellian tactics to obtain his "means" to his end. Romeo responds to the Apothecary's refrain from the exchange in fear of the death-sentencing "Mantua's law" against selling poison (5.1.66–7) by referencing a larger law conspiring against the Apothecary: "The world is not thy friend, nor the world's law; / The world affords no law to make thee rich; / Then be not poor, but break it, and take this" (5.1.72–4). Romeo's invocation of a law larger than those generated by Mantua's state machinery appears to empower the Apothecary, through the provisions provided by Romeo, to subvert the "world's law" against the poor in which Mantuan legislation participates. Although he frames the sale as a radically subversive act endowing the poor with the power to "break" the laws of the world, one must remember that Mantua is "no world" to Romeo but a world for those condemned to a purgatorial and/or hellish afterlife (3.3.17–18). Made in the malevolent "no world" of Mantua, Romeo's purchase inverts itself and becomes a sale with soul-destroying effects: "There is thy gold – worse poison to men's souls, / Doing more murder in this loathsome world / than these poor compounds that thou mayst not sell. / I sell thee poison, thou hast sold me none" (5.1.80–3). Following Romeo's logic, Romeo has paid the Apothecary's "poverty and not thy will" (5.1.76), alleviating the Apothecary of any responsibility in Romeo's death. Yet under his seeming benevolence and radicalness is a plot against the Apothecary, whom Romeo wills to die, either as the result of the Apothecary's illegal sale of poison in Mantua, which Romeo "writes" home about in his letter to Montague (5.3.287–8), or as a result of the Apothecary's inverted purchase of gold, that poisonous substance to "men's souls" in "this loathsome world" of Mantua. Neither course of death to the Apothecary really matters, for in Romeo's imagination, Mantua is "purgatory, torture, hell itself" (3.3.18), and therefore its citizens are previously deceased, condemned souls.

Like Romeo, Juliet conceives of love in financial terms, and their shared preoccupations with the business of love recall Juliet's mother's lesson in matrimony as an act of monetary endowment: "So shall you share all that he doth possess, / By having him, making yourself no less" (1.3.93–4). Unlike Romeo, Juliet's "means [are] much less" (2.P.11), and thus she carries no coinage as compensation for her servant's "pains." Nonetheless, she still commands her servant to carry on her missions, sending the Nurse on one "errand" (3.3.79) after another, all in defiance of her father/the Nurse's employer. The Nurse bemoans Juliet's labor-intensive commands, complaining, "I am the drudge, and toil in your delight" (2.5.76), yet imagines that revenge against her mistress will be undertaken on her part by Romeo, whose sexual demands will "burden" Juliet: "But you shall bear the burden soon at night" (2.5.77). Alternatively, Juliet conceives of her wedding night as the means to "high fortune" (2.5.79), not as an onerous task without reward. Awaiting her nighttime conjugal appointment with Romeo, Juliet worries that, without sexual intercourse, her "high fortune" will become a business loss: "O, I have bought the mansion of a love / But not possess'd it, and though I am sold, / Not yet enjoy'd" (3.2.26–8). Either as purchaser or purchase in the marriage contract, Juliet fears she will be cheated out of a payoff of the sale: sexual use. Romeo and Juliet's matrimonial bliss cannot be divorced from sex or finance, as marriage is, according to Juliet, a financial transaction contingent upon conjugation. Without sex, Juliet can be neither the possessor nor the possessed of "a love," as her love, invoking the business of prostitution, is securable only through contracted sexual intercourse.

Juliet's conflation of love, sex, and money undermines her earlier attempt to assert her love's worth by measure of its transcendence over material properties. Prior to her clandestine Church wedding, Juliet insists that the "worth" of her love cannot be computed: "They are but beggars that can count their worth, / But my true love is grown to such excess / I cannot sum up sum of half my wealth" (2.6.32–4). Despite her poor math skills, Juliet is able to place a monetary value on her love by her very insistence on its excessive richness. Those who can compute are "but beggars" compared to her who possesses a "wealth" of "true love" that cannot be added, even in halves. Seemingly exceeding Verona's signification system, her love's transcendent "worth" still remains expressible in terms of Verona's commerce-based official culture, as it is able to be articulated by the signifier "wealth." Moreover, as we learn after the Church nuptials, her love/"wealth" cannot be possessed without Romeo, the lover/husband/business partner/purchaser/purchase whom she bought in marriage and was sold to. Her "wealth" of love is inaccessible before marriage, locked in a treasure chest; post-marriage, it still requires an exchange of keys to be accessed and/or secured.

Lacking direct possession of monetary means until marriage, Juliet relies on Verona's religious state machinery to establish her love/"wealth." This conveyance of her love further emphasizes her investments in the ideological networks of Verona's official culture. Although recognizing her relationship with Romeo as an act of rebellion against the ideology of the families' feud, she demands that it follow the codes of conduct dictated by the state. Juliet insists that Romeo prove his love through the convention of marriage: "If that thy bent of love be honorable, / Thy purpose marriage, send me word tomorrow / By one that I'll procure to come to thee, / Where and what time thou wilt perform the rite" (2.2.143–6). That their union must be legitimized through the "rite" of the official auspices of the Church, and thus recognized under God's law, is further evidence that their relationship, although in defiance of their families, remains delimited by the state-supporting ideological structures within which they have been reared. Furthermore, the primary mediator of their love is a friar who willingly joins the pair in holy matrimony not because he believes in the sincerity of Romeo's love, but because their union could end the feud disrupting the state: "In one respect I'll thy assistant be. / For this alliance may so happy prove / To turn your households' rancor to pure love" (2.3.86–8). Juliet's decision to choose a mate rather than allow her father to assign her "[a] gentleman of noble parentage" (3.5.179) may be an act of rebellion against an earthly patriarchal order; yet, since she insists that her choice be legitimized by an official other and through conventional means, it reconfirms the conception of a hierarchy based on a heavenly father whose image is conducted through the offices of a Church whose higher purpose is to support the state.

The events leading up to Romeo and Juliet's double-suicide demonstrate their increasingly diminishing transversal wherewithal, by which they could have transcended their subjective territories and Verona's official culture. Given the opportunity to make a transversal leap, to violate the familial and Verona laws that prohibit their union, the exiled Romeo prefers instead, following the Friar's advice, to patiently wait in Mantua for "a time / To blaze your marriage, reconcile your friends, / Beg pardon of the Prince" (3.3.149–51). The "time" to remedy the situation by having their relationship validated through a public announcement and a reprieve from the Prince is curtailed by Juliet's apparent death, orchestrated, again by the Friar, to avoid her enforced marriage to Paris and quietly "free [her] from this present shame" without questioning the constructs that condemn her love as "shame[ful]" (4.1.118). The closest the characters come to breaching the parameters of their official culture is with their self-inflicted deaths, when they embrace the damnation they would have been taught awaited those who committed suicide, preferring eternity within the confines of hell to life in a world whose emotional and conceptual boundaries deny and inscribe their desires.

Friar Laurence's role in the events leading up to Romeo's and Juliet's deaths serves to remind the audience of the almost always present presence of state power in this play. This hovering, affective presence challenges official interpretations of *Romeo and Juliet* as the epitome of romantic tragedies. Friar Laurence is a relatively minor character whose pragmatic outlook on Romeo and Juliet's relationship can be read as a viewpoint included to veil the materialism of Romeo and Juliet's love by providing a juxtaposing voice of reason that, through contrast, emphasizes the lovers' rashness and emotional vitality, and thereby masks their financial preoccupations. Nevertheless, the Friar's role as an official of the Church and his desire to use their union to repair the feud disruptive to Verona demonstrates that Shakespeare's play is more about state power than love or money, per se. This less idealistic reading is confirmed by the actions of the remaining characters in the aftermath of the lovers' suicides, when Romeo and Juliet undergo their final metamorphosis and become golden.

StoryBrooke romance

In Lady Capulet's discourse on Prince Escalus' kinsman Paris lies another golden thread that ties Romeo and Juliet's love to the commercial cinderblocks of "Verona's walls" (3.3.17), and to the market for tragic love in Verona and beyond. In her promotion of Paris to her daughter, Lady Capulet articulates a formation of desire infused with the financially-driven imperatives of Verona's official culture. Asking Juliet to consider the nobleman as husband-to-be, Lady Capulet calls upon gold-bound literature, describing Paris as a "precious book of love" (1.3.87) written "with beauty's pen" (1.3.82) and enjoyed only by those possessing the golden key of upper-class status: "That book in many's eyes doth share the glory / That in gold clasps locks in the golden story" (1.3.91–2). In comparing Paris to a beautified text available to the reading public, those literate "many's eyes" who can "share the glory" of the otherwise "gold" locked "golden story," Lady Capulet reminds Juliet and the audience that Paris' readership, and the ideology of love he represents, circulates among those whom Capulet's illiterate servant refers to as "the learned" (1.2.44), that group of reading and "writing person[s]" (1.2.43) to whom, as the same servant's recourse to Romeo makes clear, Romeo belongs. Like an eager librarian, Lady Capulet impresses upon Juliet the value of perusing as well as engrossing herself in Paris' text in order to become its "cover" (1.3.88), a grafting that would give Juliet the opportunity to undergo the socially-rescriptive process of marriage in Verona, where financial endowment cannot be divorced from love: "So shall you share all that he doth possess, / By having him, making yourself no less" (1.3.93–4).

Transposing herself onto "the volume" of Paris (1.3.81) would allow Juliet to become one of Verona's "ladies of esteem" (1.3.70) and enter into Verona's royal family, becoming "no less" than Paris' social worth, which is signified by the "gold clasps" preventing his content from being accessed by anyone but the privileged few.

Juliet prefers instead to marry within her social class to one similarly schooled as she, as indicated by her chiding, "You kiss by th' book" (1.5.109), that kissing text both Romeo and Juliet are familiar with. Juliet's rejection of Paris, the noble "gentlemen" more "worthy" than herself (3.5.145) that the Capulets try to teach her to choose, suggests that she has forsworn the lesson in love, investiture, and matrimony found in Paris' "volume" (1.3.81). Nevertheless, upon hearing of Tybalt's murder by her new husband Romeo, Juliet recalls the metaphor used by her mother to instill the lesson: "Was ever book containing such vile matter / So fairly bound" (3.2.83–4). While "fairly bound" here directly refers to the "sweet flesh" of Romeo (3.2.82), it also indexes Juliet, that "fair daughter of rich Capulet" (2.3.54) taught by her mother that the role of wife was to serve as beautifying "cover" for Paris, the "unbound lover" who as husband provides social mobility and financial security (1.3.87–8). Substituting Romeo for Paris, Juliet adheres to the imperatives of matrimonial bonding in Verona, and transfers her beautifying function to the "vile matter" of the murderous book of Romeo.

Romeo and Juliet's union flouts the Paris–Capulet marriage contract, although it is imbued with the same financial opportunism, and it bears the potential to bridge an alliance between two merchant households "alike in dignity" (1.P.1). Alternatively, Paris and Juliet's wedding was capable of producing a merger between merchant class and royal family. Juliet's apparent death before betrothal to Prince Escalus' kinsman, then, functions as a preventive act securing the inherited right of nobility by disallowing the dilution of royal blood with mercantile. Intended as a measure to "chide away" the "shame" of bigamy (4.1.74), Juliet's disguise as a "thing like death" (4.1.74) instead results in the destruction of Paris. Paris' demise, accompanied by the earlier killing of Mercutio, the only other kinsman of Prince Escalus identified in the play, depletes the Prince's lineage, and yet protects his seat of power from usurpation and his bloodline from non-noble pollution.

The death of Paris also leaves an open entry in Verona's library, whose holdings also include "th' book" on kissing Juliet references (1.5.109), the "book of arithmetic" Mercutio accuses Tybalt of fighting by (3.1.103–04), and "sour misfortune's book" (5.3.82), in which both

Paris and Romeo are scripted. The now absent "precious book" of Paris/ love, which Juliet has previously used as a reference guide, is replaced with a union between mercantile class and monarchy that takes the form of a textual co-production co-opted at its conclusion by the reigning patriarch. The Prince and the feuding fathers fill in the gap generated by Paris' murder by co-authoring a new story, still golden, but more woeful and accessible to a broader public. Reimagining the lovers as golden statues, the fathers bind their children with the gold bonds of Paris' text, locking them into Verona's discourse on matrimonial love as financial investiture. As statues, Romeo and Juliet will be endowed with the static quality of words on a page, destructible only with the ruin of the material in which they are set. The audience of these static figures will not be restricted to the learned few, as many more "eyes" can "share the glory" of Romeo and Juliet's publicly displayed "golden story" (1.3.191–2) than Paris' book, which enjoyed a limited circulation as opposed to the indefinite readership envisioned for Romeo and Juliet, who will remain golden "whiles Verona by that name is known" (5.3.299). And while Verona remains known as Verona, Romeo and Juliet are to be read as officially interpreted by Prince Escalus: "For never was a story of more woe / Than this of Juliet and her Romeo" (5.3.308–09). Echoing the First and Second Act Prologues, whose function is to frame "the two hours' traffic of our stage" as a theatricalization of a past event (1.P.12), the Prince reminds the audience that *Romeo and Juliet* "was a story," a representation from the past performed in the present to secure the future posterity of Verona's official territory.

The concluding act of textual co-production can be said to reenact the play's authorial construction, especially as the Prince and Shakespeare merge into the role of official authorial agent. The plethora of references to texts and literary tropes in the play, suggestive of the circulation of ideologies as learned discourses, also implies an awareness of the growing printing industry of the late sixteenth century, with its potential to circulate an authorized text and to produce pirated copies. Anticipating reproductions and extensions of R&Jspace, the conclusion strives to reassert single-authorial presence by advertising the play's value, as literature and/or theater, over other alternative editions, including its primary source-text. Just as Prince Escalus scripts in collaboration with the feuding fathers, Shakespeare adapts from previous sources, primarily Arthur Brooke's poem *The Tragicall Historye of Romeus and Juliet* (1562). Like the Prince, it is Shakespeare who appears to emerge as the sole-authorial agent, as it is his name, along with

Verona's, that has been attached to the future reproductions of the text that the conclusion points toward. The Prince's final two lines, "For never was a story of more woe / Than this of Juliet and her Romeo" (5.3.308–09), also affirm what the title of the play's second quarto announces, perhaps in response to piracy: this is "The Most Excellent and Lamentable Tragedie of Romeo and Juliet," preferable to all other existing or foreseeable unauthorized and/or alternate versions.[24] As the Prince, striving to reaffirm his power over the populace, attempts to secure the lovers' text from alternative interpretations by insisting upon its unsurpassable tragic quality, the conclusion strives to secure for itself a place in the present and future by reimpressing upon the audience the value of its story as a "Tragedie of Romeo and Juliet" symbolized by glorified golden figures, not the marble tomb monuments Brooke promises for his lovers at the conclusion of his poem.[25] The Prince's emphasis on the play as a "story" also reminds the audience that *Romeo and Juliet* is a reproducible "story of more woe" (5.3.308), that is, it is a "Tragedie" rather than a "Tragicall Historye," and therefore Shakespeare's *Romeo and Juliet* can be displayed on page and/or stage, guaranteeing for itself a cross-class audience denied access to both the book of Paris and the textually-bound Brooke poem.[26]

The play's final attempt to reaffirm Verona's official culture in the iconicized figures of "Juliet and her Romeo" (5.3.309) celebrates the commercial opportunities, in print and/or in performance, of reproducible romantic love. Infused into the metallically-morphed lovers, love becomes less transcendental and more material, locatable by reference to monetary figures referencing a value imagined to be as stable as Verona's name. In effect, *Romeo and Juliet* laments precisely what it excels at representing by undermining, through financial subterfuge, the glorification of its woeful lovers. Because Romeo and Juliet are unable to conceive of love outside the commercial values of Verona's official culture, to express their desire without recourse to Verona's sociopolitical conductors, or meet without reliance upon an intact labor system, the romanticization of Romeo and Juliet's love as capable of conquering all crumbles into an idea of love as both corrupted by commerce and commercially sublime.

Unlike the later Troilus and Cressida, who creatively propel themselves into a future-present-space where they will be endowed with the affective presence of signs of fidelity (Troilus) and infidelity (Cressida), Romeo and Juliet can only conceptualize a place for themselves outside the confines of "Verona's walls" in the after-life promised by Christian doctrines on death: either in that "purgatory, torture, hell itself" awaiting them for committing suicide (3.3.18), or in the heavenly bliss awaiting the redeemed. The lovers' secular after-life as "pure gold" symbols of tragically thwarted love is projected by their fathers and the Prince at the conclusion. Romeo, however, does foretell of their resurrection when, comforting Juliet before leaving for Mantua, he points towards R&Jspace, including the trajectories of R&Jspace in recent criticism that we have explored: "all these woes shall serve / For sweet discourses in our times to come" (3.5.52–3). In their "times to come" thus far, Romeo and Juliet have been transposed into R&Jspace, which has become a site for scholarly struggles over the value – historical, political, social, cultural, and transcendental – of the author, the play, and its characters. As such, R&Jspace continues to push the conceptual boundaries of Shakespeare scholarship, consequently preventing the textual closure imagined by the play's Prince, and disallowing its iconic lovers to remain statically bound by golden fetters.

Future-present transversality "some other where" (1.1.295–6) in R&Jspace

Before departing, we would like to take a step back and reassess our analysis. What you have just considered is a materialist interpretation of *Romeo and Juliet*, a detailed reading of the play that provides new insights, yet is consistent in many respects with poststructuralist approaches to Shakespeare, particularly those that are also new historicist. Shakespeare's plays have in recent years become a vehicle for arguments against essentialism in favor of indeterminate selves, and it is within this framework that this analysis has thus far proceeded, revisiting a pre-existing thematic, undoubtedly with a number of surprising twists, but also with the shared aims of challenging conceptions of Shakespeare as a playwright representing and/or promoting universal humanity and absolute truths, such as an inherent, idealized, and/or universalized formation of desire. In tracing the monetary means at work in the text, we have located Romeo and Juliet's relationship as simultaneously working within and without the power networks of their subjective territories, only to, in the end, substantiate Verona's official territory. Following the play's fugitive financial elements, we have refuted idealized conceptions of Romeo and Juliet as transcendental lovers, and in the process offered a reading suggestive of the subversion-containment paradigm found in much new historicist criticism. This can be seen in our argument that Romeo and Juliet's ostensibly transgressive desire is actually encouraged by the state machinery, which usurps it so as to consolidate state power. Our intention in writing this postscript is not to debunk new historicism or the analysis you have just considered. Instead, in keeping with transversal poetics' investigative-expansive methodology, we now want to reparameterize our fugitive explorations of *Romeo and Juliet* as a means by which to reconsider the play's potential transversal effects. If the lovers' truncated transversality only functions to consolidate a monarchical order invested in capitalistic enterprise, does the play lack transversal power? In other words, if not in their love, where else may the play's transversal power be found, if it is to be found at all?

According to Davis, the structure of *Romeo and Juliet*, bookended by the First Act Prologue and Prince Escalus' closing remarks, posits the action in "a kind of liminal phase" confining the characters "between a determining past and future" (57). Also recognizing the liminality produced by the play's structure, we interpret the narrative framework as a transversal device that instead generates an indeterminable past, present, and future within and exceeding the action represented on stage. The framing of the characters and their lives as simultaneously here (on stage) and there (off stage) by the first Prologue and the Prince's last decree produces a theatricalized transition state between life and death, past and present, "real" and theatrical, worldly and other-worldly, and represented and absent that, through the act of resurrection, transports the main characters, and along with them the audience, into non-delineated spaces where corpses commingle and merge with the living.

The First Act Prologue frames the action of the play as a replay, a theatrical representation in the present of a tragedy of the past: "The fearful passage of their death-mark'd love / And the continuance of their parents' rage, / Which, but their children's end, nought could remove, / Is now the two hours' traffic of our stage" (1.P.9–12). The Prologue's announcement to the audience that the "traffic" they are "now" to witness is a presentation of the events which lead up to Romeo's and Juliet's "end" situates the play within the conventions "of our [Western] stage" that are endowed with the power to represent the revisited or imagined past in the present of a live performance.

In *The Rainbow of Desire*, Augusto Boal accounts for the transformational quality of the literal stage by describing the performance area as an "aesthetic space" that "*is* but *doesn't* exist" (20), an apparent contradiction that allows for endless possibilities: "In the aesthetic space one can be without being. Dead people are alive, the past becomes present, the future is today, duration is dissociated from time, everything is possible in the here-and-now, fiction is pure reality, and reality is fiction" (20).[27] Embracing the "plasticity" of space and time, the stage can become a portal for journeys beyond the "physical space" that exists just outside the theater, transporting the audience into a world where spatiotemporal demarcations and the socially-prescribed codes that inhibit interactions among humans and between things can collapse (Boal 20–1). It is into this "aesthetic space" that the First Act Prologue draws the audience, allowing the spectators to believe that what they see before them is the portrayal of the lives of two already dead "star-cross'd lovers" before and after they "take their life" (1.P.6).

Having introduced the title characters as previously deceased, the first Prologue asks the audience to hereafter accept them as living, that is, until "their death-mark'd love" reaches its culmination (1.P.9). The characters, "mark'd" as dead, are transported via the stage back to life, resurrected in the form of the two actors who are portraying their roles. With this in mind, Romeo's lines, "Tut, I have lost myself, I am not here. / This is not Romeo, he's some other where" (1.1.295–6), take on another meaning that is both metatheatrical and metaphysical. As framed by the First Act Prologue, the Romeo before the audience is literally "not Romeo"; rather, he is a representation of a Romeo who is "not here" because he is deceased. The switch from first person ("I have lost myself, I am not here") to second ("This is not Romeo, he's some other where") suggests a fluctuation between representational and "real": "I [the actor] have lost myself," and "This [the character] is not Romeo." The "real" Romeo, not the actor or the actor-as-character, is "some other where" outside of the representation, but he remains within its frame in what we call, borrowing from Shakespeare to respond to Boal's "aesthetic space," a "some-other-where-but-here-space." "Some-other-where-but-here-space" is an unaesthecized, absent-present affective space, a non-represented yet aesthetically-invoked dimension conjured through a performance that posits its existence, but cannot confine it to the conventions of a specific performance area; a "some-other-where" existing in a simultaneous past (Romeo's and Juliet's "real" deaths) and present (the live performance of Romeo's and Juliet's deaths).

Knowledge of the conclusion of *Romeo and Juliet* haunts the onstage representation of the main characters, leading the audience to anticipate the announced outcome, which can be viewed as a replay of their initial suicides, or a "second-coming" of their first deaths. In effect, the characters are both dead and not dead, beings without being that have come alive through a performance that culminates with their lives lost.

The actors portraying their roles are mediums for the deceased characters, channeling them back from the beyond "some-other-where" to the present spatiotemporal plane of "here." The actors become conduits for the spacetime travel of the audience, who, accepting the aesthetic conventions of the stage, are conveyed through their observation of the representational forms from the past to the present portrayal of these figures whose elusive whereabouts are referred to yet not shown.

Romeo and Juliet's onstage journey ends in the tomb as expected, marking a circular route from death to life to death. The conclusion of the play resists such closure by not returning to the present, but instead leaping to a "some-other-where" in the future. Rather than revisit the narrative device that establishes the performance as an "aesthetic" representation by attaching an epilogue, the play ends with Prince Escalus, one of the characters from within the replayed past, delivering the final lines which, as Callaghan notes, look forward to future retellings of the story (61–2). The play thus ends in the past as played in the present with an eye towards the future, blurring delineations of time as linear, and leaving the audience oscillated backwards and forwards into an aesthetic here and an offstage, unaesthetic "some-other-where" whose only tangible quality is its fictionality, reintroduced by the Prince at the conclusion: "For never was a *story* of more woe / Than this of Juliet and her Romeo" (5.3.308–09; our emphasis).

The aesthetic "here" and unaestheticized "some-other-where" of *Romeo and Juliet* challenges conventional perceptions of past, present, future, life, death, post-death, real, theatrical, and the spacetime that distinguishes the relationships among these concepts and experiences. The play's boundless terrain suggests that what appears in the now may be a replay of the then, or perhaps an allusion to a "where" that cannot be made substantial according to the perceptions of space and time employed by official cultures to situate its members along a shared spatiotemporal plane. By shifting the sands of spacetime, the play's framework undermines this endeavor, revealing the negotiability of spatiotemporal boundaries as well as its own fugitivity. In the play, spacetime becomes a gateway for possibilities closed to the onstage title characters but opened to the offstage audience who, asked to think non-linerally and to imagine death and life as simultaneous states of existence, are transported into a non-delineated "some-other-where-but-here-space" that is a transversal territory whose transversal power issues forth from its disavowal of containment within conventional constructs of dimensionality. The audience therefore journeys into a spacetime unbound by state-supporting, intangible boundaries where desire, unfettered by commerce, would not necessarily lead to a doomed conclusion.

Notes

1. "Romeo" can be found on Sublime's album *Second-Hand Smoke* (MCA, 1997), and on the soundtrack to Lloyd Kaufman's *Tromeo and Juliet* (Tromeo Entertainment, 1997).
2. The conjoining signifier of Shakespeare's title was recently shortened to a plus sign in Baz Luhrmann's 1996 film *William Shakespeare's Romeo + Juliet*, presumably because the "+" is a more romantic conjunction than "and" in contemporary teen culture. Luhrmann's mathematical mark conjures the image of the holy cross used in the most (in)famous of crucifixions, and thus his title announces as sublimely sacrificial the conclusion of the lovers' lives, each of whom are positioned on opposite vertical ends of the supplemental conjunctive.
3. While the conductors of "state power" function to produce a hegemonic culture and "image of the totalized state," as Reynolds states, the "use of the term 'state machinery' should make explicit the multifarious and discursive nature of state power, and thus prevent the misperception of this dynamic as resultant from a conspiracy led by a monolithic state" ("The Devil's House, 'or worse': Transversal Power and Antitheatrical Discourse in Early Modern England," *Theatre Journal* 49.2 (1997): 143–67; 145).
4. Harold Bloom argues that Shakespeare is now "the universal canon," and that his ability alone to retain his place in the recently critically-assaulted Western canon is evidence of his universal value as a transhistorical guide to living: "Shakespeare is not only in himself the Western canon; he has become the universal canon, perhaps the only one that can survive the current debasement of our teaching institutions, here and abroad. Every other great writer may fall away, to be replaced by the anti-elitist swamp of Cultural Studies. Shakespeare will abide" (*Shakespeare: the Invention of the Human* [New York: Riverhead, 1998], 17). Stanley Wells similarly notes that Shakespeare has managed to withstand the "attack" against "the established canon of English literature," maintaining his place as "the basis of a literary education" and "the centre of a huge academic industry" despite the onslaught of decentering trends in recent criticism (*Shakespeare: A Life in Drama* [New York: Norton, 1995], 1).
5. In "No Holes Bard: Homonormativity and the Gay and Lesbian Romance with *Romeo and Juliet*," Richard Burt examines the history of queer adaptations of *Romeo and Juliet* that have challenged heteronormative readings of the play, including "homonormative" versions that, however utopian, counter authorized sexuality "either by seeking to be unlegible as gay or by designifying gender difference" (*Shakespeare Without Class: Misappropriations of Cultural Capital*, eds. Donald Hedrick and Bryan Reynolds [New York: Palgrave – now Palgrave Macmillan, St. Martin's Press, 2000], 153–86; 157).

6. For an example of Shakespace intersecting with *Romeo and Juliet*, see Hedrick and Reynolds' discussion of *Shakespeare in Love*'s (1998) symbiotic representation of the production of Shakespeare's off- and onstage romances in "Shakespace and Transversal Power" (*Shakespeare Without Class*, 3–47).

7. While Brooke's poem is said to be the first English translation of Pierre Boasistuau's 1599 French translation of Mateo Bandello's version of the story, itself an adaptation of Luigi Da Porto's work, the tale at the heart of Brooke's poem is believed to have been derived from Western folklore. Accounting for what he deems the universal popularity of *Romeo and Juliet*, Northrop Frye argues that the play is derived "from an archetypal story of youth, love and death that is probably older than written literature itself" (*Northrop Frye on Shakespeare*, ed. Robert Sandler [New Haven: Yale University Press, 1986], 30). Wells echoes Frye when he attributes the mythic status of the play to, at least in part, the "numerous versions of the basic story" that preceded Shakespeare's adaptation (*Shakespeare*, 77). For a summary of the transmission of the story from fifteenth and sixteenth century European *novelle* to Brooke's poem to Shakespeare, see pages 32–7 of Brian Gibbons' introduction to the Arden edition of *Romeo and Juliet* (London: Methuen, 1980), 1–77.

8. The investigative-expansive (i.e.) mode is transversal theory's response to the dissective-cohesive (d.c.) mode that pervades scholarship, in which disparate elements of a given subject are pursued, often at the exclusion of interrelated aspects, to build a cohesive argument in the interest of producing a totalized conclusion. Alternatively, the i.e. mode remains open to reparameterizing as the analysis progresses, allowing the possibility for interrelated elements to re-route the analytical path. For a comprehensive guide to investigative-expansive methodology, see zooz, "Transversal Poetics: I.E. Mode" in Reynolds, *Performing Transversally: Reimagining Shakespeare and the Critical Future* (New York: Palgrave Macmillan, 2003), 287–301.

9. All citations from *Romeo and Juliet* are from the Arden edition, ed. Brian Gibbons (London: Methuen, 1980).

10. Following the bylaws of much materialist analysis of the past thirty years, particularly its investment in Williams' ideological diagram, many critics have examined clashing early modern discourses in *Romeo and Juliet*. See, for instance: Courtney Lehmann's analysis of the conflict between the "residual ideology of *auctoritas*" and the "emergent ideology of self-authorship" in the play ("Shakespeare Unauthorized: Tragedy 'by the book' in *Romeo and Juliet*," in Lehmann, *Shakespeare Remains: Theater to Film, Early Modern to Postmodern* [Ithaca: Cornell University Press, 2001], 25–53; 25); Lloyd Davis' reading of the play as "a complex intersection between historical and emergent discourses of desire" that generates a conception of desire that "helps to initiate" a later dominant literary tradition of melancholic love that extends from Shakespeare's later tragedies to F. Scott Fritzgerald's *The Great Gatsby* (" 'Death-marked love': Desire and Presence in *Romeo and Juliet*," *Shakespeare Studies* 49, ed. Stanley Wells [Cambridge: Cambridge University Press, 1996], 57–67; 67); Dympna Callaghan's analysis of how *Romeo and Juliet*, participating in the production of a post-feudal form of patriarchy, "consolidates a certain formation of desiring subjectivity

attendant upon Protestant and especially Puritan ideologies of marriage and the family required by, or at least very conducive to the emergent economic formation of, capitalism" ("The Ideology of Romantic Love: the Case of *Romeo and Juliet,*" *The Weyward Sisters: Shakespeare and Feminist Politics,* eds. Callaghan, Lorraine Helms, and Jyotsna Singh [Oxford: Blackwell, 1994], 59–101; 59); and Greg Bentley's interpretation of the text's attitude toward money as evidence of the early modern period's inheritance of the Medieval "association between pride and money, or cupidity" ("Poetics of Power: Money as Sign and Substance in *Romeo and Juliet,*" *Explorations in Renaissance Culture* 17 (1991): 145–66; 152). In reading recent criticism of the play as a space of combative discourses and interpretative strategies, we too pay homage to Williams' work. For Williams' notions of dominant, residual, and emergent cultures, see *Marxism and Literature* (Oxford: Oxford University Press, 1977), 121–8.

11. For Williams' concepts of "structures of feeling," see *Marxism and Literature,* 121–36.

12. Frye dismisses the question as to "whether Shakespeare's audience would have assumed that Romeo was damned for committing suicide" as "tedious," and argues that the audience would more likely have recognized Romeo as a conventional "lover-hero" devoted to "the religion of love" whose moral standards differ but do not "collide with Christianity" (*Frye on Shakespeare,* 24). Frye's contention that, for Shakespeare's audience, the literary rules of love, differing without opposing Christianity, would supersede those of religion is undermined by key components of the text excluded from Frye's analysis. Whether Juliet's self-chosen death would similarly have been read as a conventionalized deviation from, without denouncement of, the tenets of Christianity is not addressed by Frye, who similarly does not account for the Friar's admonishment of Romeo for considering the sin of suicide as a viable option. While the Friar denounces suicide as a "damned hate upon thyself" (3.3.117), Juliet regards it as an empowering act: "If all else fail, myself have power to die" (3.5.242). Endowing the human "self," instead of the divine, with the "power" to take away life, Juliet's declaration does "collide" with Judeo-Christian religious teachings: Juliet's pronouncement is a rejection of the omniscient God-figure of the Protestant and Catholic faiths, and it transforms suicide from a "damned hate" into evidence of self-determination. Confronted by Juliet endowed with the "power to die," the Friar is impressed by the self-will signified by her willingness to take her own life: "Thou hast the strength of will to slay thyself" (4.1.72). The contradictory codings of suicide as both a ticket to hell and an act requiring "strength of will" can be interpreted in a number of ways: Shakespeare was aware of his audience's potential religious qualms and so sought to rectify the lovers' damnation by self-chosen death through Juliet and the Friar's re-reading of suicide as a sign of "strength," not sin; Romeo's and Juliet's suicides subversively question the doctrines of the dominant and residual religious structures of the period; the play's moral guidelines regarding suicide differ according to the gender of the suicidee, as it is deemed damnable if enacted by Romeo and courageous if enacted by Juliet, and thereby indicate a preference for preserving the life of males; and/or the Friar's and

Juliet's opposing accounts of what suicide signifies emphasizes the conflicting moral differences between conventionalized discourses of love and those of Judeo-Christian religions.

13. Callaghan's reading of the text as an affirmation of what in transversal terms would be regarded as state power is not at odds with Frye's analysis of the play's political agenda in *Frye on Shakespeare*. Although Frye regards Shakespeare as apolitical and insists that his plays only reflect, and do not actively participate in, "aspects of social life" identifiable to and suited to satisfy the "assumptions" of his audiences (2), he does find in *Romeo and Juliet* a "political moral" akin to that found in the *Henriad* (15). Like the Tudor history plays, *Romeo and Juliet* offers a reaffirmation of monarchical power in the person of Prince Escalus, who emerges as the triumphal agent quelling the chaos caused by the feuding feudal fathers in the play (15). Furthermore, in light of Elizabeth I's "personal dislike of duels and brawling," the play functions to condemn that which "the authorities thoroughly disapproved of anyway" (15). Appropriating Frye's reading, we could argue that Prince Escalus is a projection of Elizabeth I, and that Escalus/Elizabeth's promise that "Some shall be pardon'd, and some punished" (5.3.307) flatters the court by celebrating as merciful yet stern the Queen's grip on her subjects. Conversely, we could argue that the play offers a veiled threat and/or advice on rulership to the offstage sovereign by implying that she, like Prince Escalus, will, "for winking at [nobles'] discords," lose "a brace of kinsmen" (5.3.293–4). These potential readings are intended to elucidate the benefits of an investigative-expansive methodology, which allows us to venture subjunctively into alternative areas of exploration as we continue with the present-space of our analysis.

14. Among critics who have identified *Romeo and Juliet* as an early, faulty work is Norman N. Holland, who argues that the text reflects Shakespeare's early writing style. To Holland, this "quintessential tragedy of opposites" (*Psychoanalysis and Shakespeare* [New York: McGraw-Hill, 1964], 331) offers a view into the "lyric, rather than dramatic" quality of Shakespeare's imagination during the period in which he was busily composing poetry (*The Shakespearean Imagination* [New York: Macmillan, 1964], 75). The text's reoccurring dichotomies, typical of "Shakespeare's early tongue," produce what Holland deems is "not a very good play" (*Imagination* 84). The text is not without redeeming use-value to Holland: not only can it be used to establish a comparative, imagined as universal, Aristotelian basis for evaluating Shakespeare's mature plays as well as "any work of art" (*Imagination* 84), but its faults have a humanizing effect on the image of Shakespeare as divine genius: "We should not set Shakespeare up as a little tin god without fault or flaw. Shakespeare is no more perfect than other men, and *Romeo and Juliet* is one of the less perfect of his plays" (*Imagination* 89). Other critics have sought to redeem this "less perfect" play by emphasizing its uniqueness in the Shakespearean canon. For instance, Susan Snyder argues that it is less an immature play than an "early experimental tragedy" in which Shakespeare, "using as a basis his string of successes with romantic comedy," creates a romantic comedy that is diverted into a tragedy by chance, that element of uncontrollable timing that in comedy works in the favor of lovers, yet here seals their doom ("Beyond Comedy: *Romeo and Juliet* and

Othello,"   William   Shakespeare's   Romeo   and   Juliet,   ed.   Harold   Bloom
[Philadelphia: Chelsea   House,   2000],   123–49;   124).   According   to   Snyder,
unlike the mature tragedies, where tragedy is generated by "both character
and circumstances, a fatal interaction between man and moment," in
Romeo and Juliet, "external fate rather than character is the principal deter-
miner of the tragic ends of the lovers" (133). While the play's lack of an
Aristotelian intertwining of "character and circumstances" can be attributed
to its experimentation with comic forms, the text still emerges from
Snyder's analysis as of lesser quality than Shakespeare's later works, such as
Othello, that "mature tragedy of love" also set in Verona which similarly
begins with comic structures but does not make the tragic mistake of dis-
associating character and fate (134). In *Shakespeare: the Invention of the
Human,* Harold Bloom likewise argues that *Romeo and Juliet* is an early play
that points towards Shakespeare's later, greater works while offering a
unique element absent from the remainder of the canon. Dismissing *Titus
Andronicus,* Bloom describes *Romeo and Juliet* as both "Shakespeare's first
authentic tragedy" (87) and as "a training ground in which Shakespeare
teaches himself remorselessness and prepares the way for his five great
tragedies" (103). While for Snyder the play's use of comic forms makes it an
experimental, albeit faulty, exception to Shakespeare's works, for Bloom, it
is its celebration of and lamentation for an inevitably doomed "unmixed
love" that makes the text unique (89). Also unlike Snyder, for Bloom, Juliet,
Mercutio, the Nurse, and Romeo, "four exuberantly realized characters,"
outreach Shakespeare's "early breakthroughs in human invention" and are
evidence of the linear trajectory of the author and his characters' full-bloom
into brilliance (89).

15. While we find Ryan's interpretation confined by its heteronormative
    assumptions, we acknowledge that our reading of his analysis is based on
    the presumption that Ryan is not trying to suggest that only pairings of
    "men and women" have yet to be "freed from the coercion" that "contin-
    ues" to prevent them from "meeting each other's emotional needs."
    Perhaps we are wrong and Ryan means to suggest that only heterosexual
    couples are trapped in a power-struggle disallowing "emotional" satisfac-
    tion. Regardless, we still find Ryan's contention that "coercion" is always
    emotionally unfulfilling problematic because it implies that sadistic, maso-
    chistic, and/or sadomasochistic expressions of desire are somehow unsatis-
    fying, or at the very least incapable of fulfilling "emotional needs."

16. Julia Kristeva comes to the opposite conclusion in "*Romeo and Juliet*: Love-
    Hatred in the Couple" (in Kristeva, *Tales of Love,* trans. Leon S. Roudiez
    [New York: Columbia UP, 1987], 209–33). According to Kristeva, the play,
    "a drama of lovehate" (218), is evidence that desire, predicated on the
    transgression of law, springs from hate and thrives on the fear of punish-
    ment from an external, third party. Marriage is "antinomic to love" (209) in
    that it legitimizes desire and thereby neuters passion. *Romeo and Juliet* also
    affords Kristeva the opportunity to put Shakespeare on the couch. Reading
    the play biographically, she regards it as a guilt-ridden, nostalgic funeral
    dirge to Shakespeare's dead son Hamnet that idealizes the playwright's
    unhappy marriage to Anne Hathaway, yet cannot veil "the hatred that
    dwells in marriage and produces death" (219). While Ryan dismisses

Kristeva's reading as "psychobiography" (118), we find it a compelling example of the intersection of Shakespace and R&Jspace. Not unlike the similarly spatially-intersecting *Shakespeare in Love*, Kristeva finds in the teenage lovers of lore the stuff from which to spruce up Shakespeare's sparse historical record, producing in the merger a biographification that extends the lovers' lives beyond the text and into the realm of imaginable realities of the daily grind of Shakespeare's life.

17. With the exception of a brief mentioning of Kristeva (93), Bloom does not name names when it comes to the "resenters" he so resents.

18. In "Poetics of Power," Bentley similarly traces the monetary moments in the play. To Bentley, these moments articulate the early modern period's paradoxical attitude towards money and function as a critique of the dehumanizing effects of "a dialectic of power" (146) by revealing money as an empty signifier whose substantive value, while imaginary, is corruptive. Taking a formalist approach to the text, Bentley argues that the play stages a "bifurcated reality" divided between "the world of politics – the feud – and the world of romance – the protagonists' love affair" (146), with each character's attitude towards money demarcating which world they belong to and characterizing the division between the "aristocratic body" and the world of the "other." Those belonging to the realm of the "other," most obviously Romeo and Juliet, "attempt to invest life with individuality and prevent themselves from being swallowed by autonomous abstractions" (148), and thus function to denaturalize the self-authorship prohibiting aristocratic system. While we find much of Bently's analysis insightful, he neglects the lovers' investment in what he defines as the "aristocratic body," and thus his reading results in a romantification of the protagonists as sacrificial victims whose deaths primarily function to denounce an aristocratic socioeconomic system forbidding self-actualization. Bentley, reading financial greed as the implied source of the feud, also does not consider the role of the state (emblematized in Prince Escalus) or the Church (symbolized by Friar Laurence) in Verona's "dialectic of power," and so neglects the power networks operating in the play against the machinations of the familial structure. In highlighting the protagonists' participation in the maintenance of Verona's official culture, we follow Bertolt Brecht's approach to the text, but with a transversal twist. In Brecht's "The Servants" (*Romeo and Juliet: Contemporary Critical Essays*, ed. R. S. White, trans. Ralph Mannheim [New York: Palgrave Macmillan, 2001], 147–51), a practice piece for actors rehearsing *Romeo and Juliet*, Brecht emphasizes the exploitation of the proletariat required for the lovers to unite, and the class privileges lurking behind the differing approaches to desire in the play. As in Shakespeare's text, where Romeo and Juliet's idealized notions of love contrast with the Nurse's and Friar Laurence's more practical approaches to desire and marriage, in Brecht's rehearsal scene, Romeo and Juliet's romanticism, characterized by declarations of willful suicide, contrasts with the notions of their respective servants, Gobbo and Nerida, whose understandings of familial responsibilities and love are based on practical considerations, such as feeding their loved ones, that are absent from the bourgeois experiences of Romeo and Juliet. Despite the title characters' self-sacrificial declarations, it is the lovers' servants who are required to sacrifice their

needs in order for the landlord Romeo and landed gentry Juliet to unite. Like Brecht, we find that the lovers' are implicated in the exploitative practices of the socioeconomic system represented in the play. Informed by transversal theory, we are also able to imagine Romeo and Juliet as exemplars of truncated transversality who, too inscribed by the boundaries of their subjective territory to transcend its mappings, are ultimately morphed into products of gold generated and securing Verona's state machinery.

19. In arguing that the "educational apparatus" has "replaced in its functions the previously dominant ideological State apparatus, the Church," Althusser emphasizes the role of the family in maintaining whichever state apparatus is prevailing: "One might even add: the School-Family couple has replaced the Church-Family couple" ("Ideology and Ideological State Apparatuses [Notes Towards an Investigation]," in Althusser, *Lenin and Philosophy and Other Essays*, trans. Ben Brewster [New York: Monthly Review Press, 1971], 126–87; 153–4).

20. Wells dismisses readings, such as ours, that deny "the qualified happiness of the play's ending" and find "a hollow materialism" in the concluding projection of the lovers into gold statues (*Shakespeare*, 82). For interpretations of the play's ending that, like ours, question the reconciliatory effect of the iconization of the lovers in the final scene, see, among others, Bentley's "Poetics of Power" and Susan Snyder's "Ideology and the Feud in *Romeo and Juliet*," *Shakespeare Survey 49: Romeo and Juliet and its Afterlife*, ed. Stanley Wells (Cambridge: Cambridge University Press, 1996), 87–96. Both Bentley and Snyder note that the competitive drive between the two fathers over the construction of the lovers into gold figures implies that, despite the Prince's public pronouncement of peace, the rivalry will continue (Bentley 163–4; Synder 96). However, the Prince's public promise of penal recourse for the deaths of Romeo and Juliet – "Some shall be pardon'd, / and some punished" (5.3.307) – suggests that any future outbreaks of disorder will not be tolerated by the state.

21. See also Reynolds' "Devil's House." Expanding upon Gilles Deleuze and Félix Guattari's term "becoming," Reynolds describes "becomings-other" as "a desiring process by which all things (energies, ideas, people, societies) change into something different from what they are; and if those things were, before their becoming, identified, standardized or normalized by some dominant force (state law, Church credo, official language), which is almost inevitably the case, then any change whatsoever is, in fact, a becoming-other. The metamorphosis of becoming-other-social-identities trespasses, confuses, and moves beyond the concepts of negation, essentialism, normality, constancy, homogeneity, and eternality, that are fundamental to subjective territory" ("Devil's House" 150–1).

22. The Nurse, having followed Juliet's orders to "give this ring to my true knight" after Tybalt's death (3.2.142), has given Romeo a ring from Juliet (3.3.162), yet the text does not indicate whether Romeo returns the favor. The sought after "precious ring" may be a metaphor for Juliet, as upon seeing her at the Capulet feast, Romeo gives her the following ringing endorsement: "What lady's that which doth enrich the hand / Of yonder knight?" (1.5.41). As Romeo is the "true knight" who eventually receives

Juliet's ring and her hand in marriage, presumably the "precious ring" he intends to steal from Juliet's corpse is actually her body.

23. We are presuming that the "that" granting prosperity that Romeo gives to Balthazar is not Romeo's letter to his father, since Romeo's earlier line, "Hold, take this letter" (5.3.22), suggest that Balthazar has already received Romeo's mail.

24. The First Quarto (1597) title page announces the play as "An Excellent and conceited Tragedie of Romeo and Iuliet," while the Second Quarto (1599) advertises itself as "The Most Excellent and Lamentable Tragedie of Romeo and Iuliet, newly corrected, augmented, and amended." According to Gibbons, the second title suggests that Q2 was intended as a replacement for the pirated Q1 (1). For a discussion of the differences between all five quartos of *Romeo and Juliet*, see, among other works, Gibbons' introduction to the Arden edition of the play, 1–23.

25. See Arthur Brooke, *The Tragicall Historye of Romeus and Juliet* in *Narrative and Dramatic Sources of Shakespeare*, ed. Geoffrey Bullough, Vol. 1 (London: Routledge and Kegan Paul, 1957), 284–363.

26. Here we follow and depart from Lehmann's argument that the play stages a struggle over authorial command. While Lehmann finds in Romeo's failed attempts to author his own destiny and transcend the fate prescribed to him in Brooke's poem a displacement of Shakespeare's "struggle between the proper imitation of his source-text and his burgeoning 'bibliographic ego' " ("Shakespeare Unauthorized," 28), we find that in the conclusion, Shakespeare's image as primary authorial agent is asserted in the person of the Prince. We agree with Lehmann that the five extant quartos of the play disallow "one authorized version" easily attributable to Shakespeare (53), but find this challenge to the authorial aspirations of the text related to the play's consciousness of the possibility of the manifestation of alternative versions. Furthermore, the quintet of quartos testifies to the popularity the Second Quarto title and the conclusion of the play imagines for itself in the realm of print culture. In noting the merger of Prince Escalus and Shakespeare at the conclusion of the play, we also depart from David Scott Kastan's argument regarding Shakespeare's lack of literary aspirations in *Shakespeare After Theory* (New York: Routledge, 1999). According to Kastan, it is evident by Shakespeare's non-involvement in the process of printing any of his plays that "[p]erformance was the only form of publication he sought for them" (72). Kastan contends that the absence of Shakespeare's name on the title page of the first three quartos of *Romeo and Juliet* reaffirms Shakespeare's disinterest in the process of printing his plays (80), and thereby his indifference to the image of Shakespeare as author the editors and publishers of the First Folio constructed (81–92). While we cannot determine whether or not Shakespeare desired the literary status he later received, we find that the posterity and prosperity of *Romeo and Juliet* in print and performance is asserted by Prince Escalus in the play's finale. Unlike Kastan, we do not delimit the imposition of an authorial presence to the printed page; rather, we recognize that the ongoing construction of Shakespeare as primary authorial figure is a negotiation requiring the presence of a receiving and believing audience. Moreover, if Kastan is correct and "[p]erformance was the only

form of publication [Shakespeare] sought for [his plays]," then it is fitting that his authority as author be asserted theatrically within the world of the play. An intriguing area for further expansion would be to consider how the struggles to attribute and/or differentiate the quartos to and/or from Shakespeare's pen engenders an intersection of Shakespace with R&Jspace by challenging and/or insuring that *Romeo and Juliet* remains Shakespeare's *Romeo and Juliet* while Shakespeare's texts by that name are known.

27. While Boal's definition of the "aesthetic space" is useful to this study, we do not subscribe to his reading of *Romeo and Juliet*. Elaborating on his "image of the rainbow of desire," Boal writes, "No sensation, emotion or desire exists in a pure state in the human being. Even that love so pure between Romeo and Juliet is not exempt from aggression or resentment" (*The Rainbow of Desire: the Boal Method of Theatre and Therapy*, trans. Adrian Jackson [London: Routledge, 1995], 150). "[T]hat love so pure" in Shakespeare's play has been constructed as such, and Boal's statement suggests that he, consciously or not, agrees with official cultural usages of Shakespeare's text as symbolic of a concept of romantic love that has, as Callaghan notes, "oppressive effects" ("Ideology of Romantic Love," 60).

Works cited

Althusser, Louis. "Ideology and Ideological State Apparatuses (Notes towards an Investigation)." *Lenin and Philosophy and Other Essays*. Trans. Ben Brewster. New York: Monthly Review Press, 1971. 127–86.

Belsey, Catherine. "The Name of the Rose in *Romeo and Juliet*." *Romeo and Juliet: Contemporary Critical Essays*. Ed. R. S. White. New York: Palgrave Macmillan, 2001. 47–67.

Bentley, Greg. "Poetics of Power: Money as Sign and Substance in *Romeo and Juliet*." *Explorations in Renaissance Culture* 17 (1991): 145–66.

Bloom, Harold. *Shakespeare: the Invention of the Human*. New York: Riverhead, 1998.

Boal, Augusto. *The Rainbow of Desire: the Boal Method of Theatre and Therapy*. Trans. Adrian Jackson. London: Routledge, 1995.

Brecht, Bertolt. "The Servants." Trans. Ralph Mannheim. *Romeo and Juliet: Contemporary Critical Essays*. Ed. R. S. White. New York: Palgrave Macmillan, 2001.

Brooke, Arthur. "The Tragicall Historye of Romeus and Juliet." *Narrative and Dramatic Sources of Shakespeare*. Ed. Geoffrey Bullough. Vol. 1. London: Routledge and Kegan Paul, 1957. 284–363.

Burt, Richard. "No Holes Bard: Homonormativity and the Gay and Lesbian Romance with *Romeo and Juliet*." *Shakespeare Without Class: Misappropriations of Cultural Capital*. Eds. Donald Hedrick and Bryan Reynolds. New York: St. Martin's Press, 2000. 153–86.

Callaghan, Dympna. "The Ideology of Romantic Love: the Case of *Romeo and Juliet*." *The Weyward Sisters: Shakespeare and Feminist Politics*. Eds. Dympna

Callaghan, Lorraine Helms, and Jyotsna Singh. Oxford: Blackwell, 1994. 59–101.

Davis, Lloyd. " 'Death-marked love': Desire and Presence in *Romeo and Juliet.*" *Shakespeare Survey 49: Romeo and Juliet and its Afterlife.* Ed. Stanley Wells. Cambridge: Cambridge University Press, 1996. 57–67.

Frye, Northrop. *Northrop Frye on Shakespeare.* Ed. Robert Sandler. New Haven: Yale University Press, 1986.

Gibbons, Brian. Introduction. *Romeo and Juliet.* By William Shakespeare. Ed. Brian Gibbons. London: Metheun, 1980. 1–77.

Goldberg, Jonathan. "*Romeo and Juliet's* Open R's." *Queering the Renaissance.* Ed. Jonathan Goldberg. Durham: Duke University Press, 1994. 218–35.

Hedrick, Donald, and Bryan Reynolds. "Shakespace and Transversal Power." *Shakespeare Without Class: Misappropriations of Cultural Capital.* Eds. Donald Hedrick and Bryan Reynolds. New York: Palgrave Macmillan, 2000. 3–47.

Holland, Norman N. *Psychoanalysis and Shakespeare.* New York: McGraw-Hill, 1964.

—— *The Shakespearean Imagination.* New York: Macmillan, 1964.

Kastan, David Scott. *Shakespeare After Theory.* New York: Routledge, 1999.

Kristeva, Julia. "Romeo and Juliet: Love-Hatred in the Couple." *Tales of Love.* Trans.

Leon S. Roudiez. New York: Columbia University Press, 1987. 209–33.

Lehmann, Courtney. *Shakespeare Remains: Theater to Film, Early Modern to Postmodern.* Ithaca: Cornell University Press, 2001.

Reynolds, Bryan. "The Devil's House, 'or worse': Transversal Power and Antitheatrical Discourse in Early Modern England." *Theatre Journal* 49.2 (1997): 143–67.

—— *Becoming Criminal: Transversal Performance and Cultural Dissidence in Early Modern England.* Baltimore: Johns Hopkins University Press, 2002.

—— *Performing Transversally: Reimagining Shakespeare and the Critical Future.* New York: Palgrave Macmillan, 2003.

Ryan, Kiernan. "The Murdering Word." *Romeo and Juliet: Contemporary Critical Essays.* Ed. R. S. White. New York: Palgrave Macmillan, 2001. 116–28.

Shakespeare. *Romeo and Juliet.* Ed. Brian Gibbons. London: Arden, 1980.

Shakespeare in Love. Dir. John Madden, Jr. Miramax, 1998.

Snyder, Susan. "Ideology and the Feud in *Romeo and Juliet.*" *Shakespeare Survey 49: Romeo and Juliet and its Afterlife.* Ed. Stanley Wells. Cambridge: Cambridge University Press, 1996. 87–96.

—— "Beyond Comedy: *Romeo and Juliet* and *Othello.*" *William Shakespeare's Romeo and Juliet.* Ed. Harold Bloom. Philadelphia: Chelsea House, 2000. 123–49.

Sublime. "Romeo." Rec. 1988. *Second-Hand Smoke.* Gasoline Alley, 1997.

Tromeo and Juliet. Dir. Lloyd Kaufman. Tromeo Entertainment, 1997.

Wells, Stanley. *Shakespeare: a Life in Drama.* New York: Norton, 1995.

William Shakespeare's Romeo + Juliet. Dir. Baz Luhrmann. Twentieth Century Fox, 1996.

Williams, Raymond. *Marxism and Literature.* Oxford: Oxford University Press, 1977.

7
Viewing Antitheatricality: or, *Tamburlaine's* Post-Theater

Bryan Reynolds & Ayanna Thompson

Metatheatricality is a term often applied to Marlowe's 1587 play *Tamburlaine*. Critics reference the costuming and language of Marlowe's lead character and argue that he was constructed through metatheatrical and metapoetic self-fashioning.[1] Alternatively, we see *Tamburlaine* as an extraordinary example of antitheatricality, one that offers its own theory of theater, what we call "post-theater," that is radically different from the naturalistic mode that came to dominate the early modern English stage. Anticipating much fringe theater and performance art today, *Tamburlaine* posits a theatrical mode that is unrealistic and non-representational, a theater that exceeds typical antitheatrical prejudices by challenging their tenets (that theater sways actors and audiences into wickedness) and moves beyond a mere metacommentary on theater. Marlowe's post-theater surfaced alternatively to the emerging form of Elizabethan theater with a self-consciousness that evades easy categorical determination; it appropriates antitheatrical terms/perceptions, defies conventional theater properties (words, action, spectacle, and naturalistic acting), and ironizes the concept of metatheater.

In Elizabethan England, Marlowe's post-theater would have been transversal with respect to the emerging generic theatrical traditions of the historical moment inasmuch as its challenge to them threatened subjective and official territories and the conventions they created and supported. It would have been anti-transversal insofar as it might have worked against opportunities for subjunctivity, empathy, projection, and transference (see Chapter 1). This could have happened by estranging or defamiliarizing audience members, unless alienation occurred in a Brechtian fashion, which would have provoked political responses and the questioning of dominant ideologies; it may have been transversal in this respect. We are skeptical of recent political readings that see the play as commentary on the supposed power or lack of power of early modern England's government, especially those that interpret *Tamburlaine I* through analyses of *Tamburlaine II*. Looking backward from part II, critics make claims about the rise and fall of Tamburlaine's power, but they almost never address part I independently. Just as *Tamburlaine* was fugitive to emerging dominant theatrical discourses with its post-theatricality, a reading of *Tamburlaine* as post-theater is fugitive to the play's critical tradition, particularly as it suggests more agency on the part of the author without losing sight of the play's complex status as a sociopolitical conductor vis-à-vis the various articulatory spaces through which it has moved and continues to move.

Transversal poetics allow us to consider *Tamburlaine* in light of transversal territory, a non-subjectified region of an individual's or group's conceptual range, such as of the conceptuality delineated by the play's critical history. Thus our fugitive exploration examines how *Tamburlaine* defies generic authorities, taking into account agency where agency has so often been neglected or evacuated by scholarship on the play. Readings that treat texts and individuals as primarily political entities often dismiss possibilities for interconnectedness, agency, freedom, and creativity. The idea of a non-subjectified region allows critics to read literature as engaging more than the political elements of a society, arguing instead that individuals, however reflected in literature, can negotiate and permeate the boundaries of a society's aesthetic parameters. In reading *Tamburlaine*, one way in which the prospect of transversal territory makes possible an alternative model is through consideration of Marlowe's appropriation of antitheatricality in the interest of developing a new theatrical genre.

Theatrical properties considered

Emphasizing the strained relationships among words, spectacle, and action, *Tamburlaine*'s prologue establishes the antitheatrical terms by which the play should be read, performed, and viewed:

> From jigging veins of rhyming mother wits,
> And such conceits as clownage keeps in pay,
> We'll lead you to the stately tent of war,
> Where you shall hear the Scythian Tamburlaine
> Threat'ning the world with high astounding terms
> And scourging kingdoms with his conquering sword.

> View but his picture in this tragic glass,
> And then applaud his fortunes as you please.[2]

While many critics have argued that, in these lines, Marlowe attempts to separate his play from the routine of the "rhyming mother wits," what the lines refer to remains unclear.[3] Although it is possible to see Marlowe distinguishing himself from hack poets, it is also possible to see him including himself and his poetry within this aping lot. The lines can be read both as, "Despite the fact that recent plays have been senseless, I will lead you to something different with *Tamburlaine*," and contradictorily as, "Despite the fact that all plays are senseless, including my own, this is where I must begin to relate *Tamburlaine*." The latter interpretation allows Marlowe to celebrate the artifice of theater notwithstanding his view of theater as the realm of "conceits" and "clownage."

The prologue continues by informing us that we shall hear the world's scourge, "Threat'ning ... with high astounding terms." If one accepts the traditional interpretation that Marlowe is singling himself out from the crowd of "mother wits," the difficulty of escaping the old "conceits" still haunts the assertions about Tamburlaine's verbal prowess. Certainly Tamburlaine's "high astounding terms" are not enough to conquer the world if he must resort to his "conquering sword" in the end. The sudden reassertion of the power of language rings hollow in light of the need for action. The force of the "conquering sword," the brute reality of action, nevertheless, is not left intact as Marlowe concludes by declaring the importance of witnessing this action. It is not enough to perform the action; someone must view it and judge its worthiness as a deed. But then the idea that Tamburlaine's actions are viewed in a "tragic glass" serves to distance them even further.

The "glass" in early modern England served as an emblem for self-inspection through self-reflection, but the notion that the reflection could never become one and the same with the thing itself was always present. In addition, the didactic imprint of the glass separates reality from its reflection by stressing the specificity of the site of learning. If reality were the same as the "tragic glass," we would not need the glass at all because the didactic lessons would be readily apparent. Marlowe calls into question the ability of the glass to reflect and convey a readily apparent lesson with the invitation to his audience to applaud Tamburlaine's fortunes as they please. Although some critics have argued that the play "repeatedly teases its audience with the form of a cautionary tale, only to violate the convention," the convention itself has ambiguity written into it.[4] The early modern glass reflects both what one should be and what one is wrongly; consider *The Mirror for Magistrates* as reflecting a positive image and *A Mirrour of Monsters*, the antitheatrical tract penned by William Rankins the same year *Tamburlaine* debuted, as reflecting a negative one – that which should be avoided.[5] To adapt Marlowe's famous lines from *Doctor Faustus*, "Was this the mirror that reflected the ideal / And taught the tortured to see Ilium"? Or "Was this the mirror that showed the horror / And taught the tortured to fear Tartarus"?[6] Viewing, then, whether theater or something else, is not as simple as it might seem, especially when a "tragic glass" is involved.

The play itself goes on to reinforce and then destabilize the importance of the theatrical properties of words, action, and spectacle. *Tamburlaine* obsessively rehearses and repeats the items set forth in the prologue – from the significance of owning language and requiring declarations to reinforcing language with action to viewing the action. For example, the character Tamburlaine stresses the importance of declarations as speech acts when he requires the Persian lords to declare that he shall reign. Having just defeated them in battle, Tamburlaine nonetheless requires their spoken declaration of consent, in effect highlighting the active force of words:

> TAMB: Theridamas, Techelles, and the rest,
> Who think you now is king of Persia?
> ALL: Tamburlaine! Tamburlaine!
> TAMB: Though Mars himself, the angry god of arms,
> And all the earthly potentates conspire
> To dispossess me of this diadem,
> Yet will I wear it in despite of them,
> As great commander of this Eastern world,
> If you but say that Tamburlaine shall reign.
> ALL: Long live Tamburlaine, and reign in Asia!
> (2.7.55–64)

Just before this dialogue, Cosroe dies at Tamburlaine's feet, clearing the way for Tamburlaine to rule Persia and Asia, yet Tamburlaine twice requires the defeated to declare him victor ("Who think you now is king of Persia" and "say that Tamburlaine shall reign"). Tamburlaine's words suggest that the act of defeating his opponents is incomplete without the verbal affirmation of his ascension: he can only withstand

Mars's onslaught with the condition of their declaration (thus, the "If" clause in line 63); what is perceived as real (Tamburlaine's conquest) must be constituted through utterances (thus, the subjunctive voice indicated by the "If" clause).

But, as in the prologue, the theatrical properties of words, speeches, and declarations are not allowed to stand securely within the body of the play. Instead, other theatrical properties continue to compete for power. Actions are repeatedly referenced as support for words; as Techelles states, "Our swords shall play the orator for us," emphasizing the ability of brute force to speak volumes in political disputes (1.2.132). The play suggests that words eventually must be left aside for the more active signification process of displaying and employing weapons. To give another example, the Median lord Agydas admits that actions reveal more than words can ever convey when Tamburlaine sends Techelles to kill him. Looking upon Techelles, who stands bearing "a naked dagger," Agydas declares:

> He needed not with words confirm my fear,
> For words are vain where working tools present
> The naked action of my threatened end.
> It says, Agydas, thou shalt surely die.
>
> (3.2.92–5)

In both examples, actions are described in linguistic terms: both Techelles and Agydas declare that the swords have supplanted words because they provide a more powerful (literally and figuratively) signification process, what Agydas calls "working tools." The meanings are as readily apparent as any spoken words, but the force is so much greater as to render words disempowered and obsolete. While not having access to words is shown to make one weak (like Mycetes' portrayal at the beginning of the play), not having a naked dagger or sword to speak for oneself is worse. This is why Agydas chooses to stab himself instead of waiting for Techelles to fulfill Tamburlaine's execution order. Seeing Agydas's dead body, Usumcasane admires his willingness to act on his own behalf: "it was manly done" (3.2.109). In this signification process, action affirms one's masculinity: it demonstrates that one not only possesses "working tools," but also that one has control over them.

When characters accept the apparently stronger and more effective signification process of action, the actions themselves become worthless if there is no audience to view the events. Although Tamburlaine and his increasing number of followers ravage the world with their weapons, Tamburlaine will not be content until every man views him as an emperor:

> But since they measure our deserts so mean,
> That in conceit bear empires on our spears,
> Affecting thoughts coequal with the clouds,
> They shall be kept our forced followers
> Till with their eyes they view us emperors.
>
> (1.2.63–7)

Tamburlaine indicates that actions are not enough to garner power, respect, and influence others. Once again he articulates his conquest in surprisingly conditional terms in light of the completed action of conquest: the conquest will not be achieved, Tamburlaine suggests, until the defeated "view" him as emperor. It is as though Tamburlaine shares the early modern English antitheatricalist perspective on the transversal power of theater. For the antitheatricalists, "Transversal power is thus transmitted visually: seeing is infectious and instigates becomings. It is also transmitted aurally and physically in the theater, *as* and *through* sound waves and resonance" (Reynolds, *Becoming Criminal* 146). Whereas Tamburlaine and his men "bear empires on [their] spears," there will be those who are unconvinced until "with their eyes they view [them as] emperors." Although the term "view" here could literally mean "understand," the theatricality of the term, especially when linked with the invitation in the prologue to "View but his picture in this tragic glass," emphasizes the theatrical intent. In other words, the old notion that "seeing is believing," even if what is seen is an image constructed with words, is crucial to the theory of theater *Tamburlaine* offers. Marlowe gestures towards the phenomenology of seeing as an experiential empiricism, a semiotic communication with profound cognitive qualities, that is the most meaningful, interpretable, understandable, and translatable. Seeing is theatrical, a vital theatrical property comprising a dynamic performance and spectacle of power and will.

The many lengthy poetic descriptions of Tamburlaine's appearance further testify to the play's post-theater enterprise. On his way to battle Tamburlaine, Cosroe stops to ask, "What stature wields he, and what personage" (2.1.6), to which Menaphon waxes with impassioned eloquence:

Of stature tall, and straightly fashioned,
Like his desire, lift upwards and divine,
So large of limbs, his joints so strongly knit,
Such breadth of shoulders as might mainly bear
Old Atlas' burden; 'twixt his manly pitch
A pearl more worth than all the world is placed
Wherein, by curious sovereignty of art,
Are fixed his piercing instruments of sight,
Whose fiery circles bear encompassed
A heaven of heavenly bodies in their spheres
That guides his steps and actions to the throne
Where honor sits invested royally.
Pale of complexion, wrought in him with passion,
Thirsting with sovereignty and love of arms,
His lofty brows in folds do figure death,
And in their smoothness, amity and life.
About them hangs a knot of amber hair
Wrapped in curls, as fierce Achilles' was,
On which the breath of heaven delights to play,
Making it dance with wanton majesty.
His arms and fingers long and sinewy,
Betokening valour and excess of strength;
In every part proportion'd like the man
Should make the world subdued to Tamburlaine.

(2.1.7–30)

It is fascinating that this laudatory description of Tamburlaine as a menacing threat comes from his adversaries before they engage in battle with him. Menaphon and Cosroe understand the message (Tamburlaine looks like the man, "Should make the world subdued"), and yet they seem powerless to stop their own destruction at his hands. The spectacle described literally speaks to them and draws them into the theatricality of the scene: they are incapable of escaping it.

Evident from these examples, however, is that the stakes keep changing and the once rationalized ground becomes increasingly groundless. When at first the power of words is emphasized with both Mycetes' horrific lack of them and Tamburlaine's amazing control of them, words are then shown to be hollow if they are not backed up by something else (as Techelles states to Tamburlaine, "Won with thy words and conquered with thy looks" [1.2.228]). When appearance is stressed with the long panegyrics about Tamburlaine's capacity to convince with his looks, the power of actions is emphasized over looks. Finally, when it appears that action is what ultimately secures strength and wherewithal, it becomes clear that *Tamburlaine* is a play in which almost no action takes place on stage. Of the five battles described, not one is indicated by the stage directions; instead, they all occur between scenes of dialogue (for example, between 2.3 and 2.4 or 2.6 and 2.7) or offstage during other scenes of dialogue (for example, in 3.3 and 5.1).[7] Like when Tamburlaine requires the defeated to declare him victor with spoken words, actions are subsumed by words; the audience hears about rather than witnesses battles.

The scene in which the Damascene virgins offer themselves as doves of peace to Tamburlaine exemplifies in miniature the ways by which theatrical properties are shown to be ineffective. Once again, words, spectacles, and actions are rendered powerless one by one. As Tamburlaine prepares to annihilate Damascus, the Governor and virgins of the town plan to make a supplication – discussed by the Governor as an impending performance:

> Therefore, for these our harmless virgins' sakes,
> Whose honors and whose lives rely on him,
> Let us have hope that their unspotted prayers,
> Their blubbered cheeks, and hearty humble moans
> Will melt his fury into some remorse,
> And use us like a loving conqueror.
>
> (5.1.18–23)

The virgins will appear crying, with "blubbered cheeks" and audible "moans." In support of this scripted plan, the Second Virgin prays that the drama will pierce Tamburlaine's heart through his "eyes and ears" (5.1.53). The theatrical properties of language, spectacle, and action are deemed the most effective means by which to achieve the town's freedom:

> Then here before the majesty of heaven
> And holy patrons of Egyptia,
> With knees and hearts submissive we entreat
> Grace to our words and pity to our looks,
> That this device may prove propitious,
> And through the eyes and ears of Tamburlaine
> Convey events of mercy to his heart.
> Grant that these signs of victory we yield

> May bind the temples of his conquering head
> To hide the folded furrows of his brows
> And shadow his displeased countenance
> With happy looks of ruth and lenity.
>
> (5.1.48–59)

In this scene both the virgins and the Governor plan to appeal to Tamburlaine through a moment of theatrical artifice, what the Second Virgin explicitly refers to as a "device" (5.1.52). They believe that when Tamburlaine sees and hears their supplication, he will necessarily experience a transversal moment of empathy: he will experience their pain, grief, and terror as if it were his own because of their dramatic reenactment of these emotions. The Second Virgin straightforwardly states her belief in the power of theater and acting to "Convey" a specific and readily identifiable message, the need for mercy (5.1.54). But, of course, Tamburlaine is immune to empathy; he is uninterested in seeing the world through the eyes of others; he is unwilling to become different conceptually and emotionally; he seems incapable of transversality. The properties of theater appear unable to influence Tamburlaine. In turn, the audience of this "two hours traffic" are exposed to the radical notion, perhaps an inverted antitheatricality, that theater is incapable of affecting anything, not empathy, not projection, not transference, not change, not even a clear or predictable meaning. Whereas this seems to work against a discovery of transversalist work on early modern English theater – that the theater was a radical site because it allowed individuals to imagine, experience, and become different things from what they are accustomed to – Marlowe's appropriation and rewriting of contemporary antitheatrical terms offers an alternative manifestation of transversalism: Marlowe's theater was revolutionary because it ironically argued for the total inefficacy of theater; it was post-theatrical.

This is not to suggest that the character Tamburlaine does not employ metatheatrical tactics. Although his actions demonstrate the ineffectiveness of metatheatricality as sociopolitical strategy, his actions replicate the process. Tamburlaine's expression and presentation of these impotent theatrical properties become, in effect, post-theater in the process. From doffing his "shepherds weeds," to displaying the variously colored tents to the unyielding Damascenes, to forcing Bajazeth to watch him eat while he attempts to starve himself to death, Tamburlaine exhibits an interest in the metatheatrical aspects of power even though they do not become the ethos of the world Marlowe created for him. He displays an acute awareness of the importance of the theatrical properties of words, spectacles, and actions; he delivers several speeches about the performativity of his own words ("I speak it, and my words are oracles" [3.3.102]); he pays careful attention to maintain his "fiery looks" when meeting his opponents for the first time; and he backs up his words and looks with equally fierce actions in battle. But it is precisely because of these gestures toward metatheatricality and his awareness of performativity that allow *Tamburlaine's* exploration of the limits of antitheatricality, metatheatrical appeals, and performative will-to-power to transcend the then emergent conventional theater.

The character Tamburlaine emphasizes this transcendence through his repeated admission and demonstration of the inefficacy of metatheater even as he employs it to great effect. It is this perversity of theatrical concept realized ironically via Tamburlaine's insistence on performance that makes *Tamburlaine* potentially a *tour de force* in post-theater, and thus a transversal theater in relation to both the emergent theatrical and antitheatrical modes of the Elizabethan period. Moreover, the ironies of *Tamburlaine's* post-theater cast questionable light on Marlowe's intentions, given that he undermines the purpose of playing throughout *Tamburlaine* while at the same time capriciously offering more theater, including a sequel to *Tamburlaine* a year later in 1588. This is not to suggest that Tamburlaine's ultimate battle, which he of course must lose, comes against the author who has inscribed him into being.[8] Instead, the fugitive elements – Marlowe, his play, and the character Tamburlaine – seem strangely in cahoots, as if wanting to show that it is the staged inaction that the audience comes to see; or that by seeing it, the audience might move subjunctively – considering "as ifs" and "what ifs" – outside of their subjective territories through the acknowledgment of the inefficacy relative to their own passivity vis-à-vis the spectacular, like in the case of Menaphon and Cosroe.

Beyond the emerging conventional theater

Tamburlaine's onstage traffic is remarkably trafficless. Onstage there are no wars, no battles, and no fights (not the wars with the Persians, not the battle with the Turks, not the execution of the virgins, and not the slaughter of the pan-African forces). In fact, the only action the audience witnesses is the brainings of Bajazeth and his wife Zabina, and this happens when no other characters are present onstage. What we call the "witness-function" is enhanced in this instance, since the audience cannot recall the event for the ears of the characters, the effect of which would bring the fourth wall curiously crashing down on the stage (given the extent to which this convention was subscribed to during the period). Nevertheless, the absence of onstage witnesses suggests audience intervention, encouraging the audience to situate themselves within the world of the play. When Zenocrate discovers the lifeless bodies of Bajazeth and Zabina, she attempts a type of empathetic response, but only affirms the powerlessness of this emotive state by parroting the line, "Behold the Turk and his great empress" three times (5.1.354, 357, 362). Zenocrate wants to believe that seeing their lifeless bodies will impart a readily identifiable meaning to all who view them, in contrast to a pursuit of "earthly pomp" in the form of "sceptres" and "slippery crowns" which will lead only to destruction (5.1.353, 356).

But hers is not the message of *Tamburlaine*. Rather, Marlowe enacts his seeming capriciousness to the end with Tamburlaine and Zenocrate crowned King and Queen of Persia. Zenocrate's final words in the play affirm that the lesson she so desperately wanted to learn from the sight of the brained Bajazeth and Zabina has already been lost. Asked if she will consent to marry Tamburlaine and become Queen of Persia, she replies: "Else should I much forget myself, my lord" (5.1.500). Forgetting seems to be the only thing guaranteed from Marlowe's theater. In this, the play suggests the audience may forget themselves, although, perhaps, only temporarily. The suggestion of temporary memory loss could challenge the antitheatricalists' claims: if theatrical (or even metatheatrical) speeches, actions, and/or spectacles cannot produce anything but merely fleeting amnesia, then the dangerous effects of theater (as the antitheatricalists believed them to be) are not certain. Forgetting is primary to theater, as implied by Anthony Kubiak and Bryan Reynolds: "consciousness depends on theatricality, the continual framing, unframing, and reframing of performances, such as of remembering and of forgetting that one is acting in a particular space-time; and theater, structurally, reflects consciousness, with its multiple framings and vistas of experience" (see Chapter 3). The play suggests that audiences may forget themselves and experience "paused consciousness," which is when one forgets, however ephemerally, where one is in space as well as time (such as when we sort of snap out of a dream when the play is over). According to transversal theory, this suspended and undifferentiated conceptual-emotional realm – a kind of transversal territory – facilitates movement outside of subjective territory as well as spurs becomings- and comings-to-be other which can occur when one is exposed to alternative realities.

Deceptively, perhaps strategically, *Tamburlaine* enacts the powerlessness of theater only to produce a post-theater that was powerfully entertaining and successful. Despite the fact that staple elements of early modern theater are rendered limp, empty, and weak within the world of the play, audiences demanded more. Although Shakespeare's *Hamlet* warns against it, Marlowe's *Tamburlaine* "out-Herods Herod" while, it seems, cleverly admitting problems with the genre. *Tamburlaine* is not naturalistic, but not because it fails at naturalism. To our knowledge this is a genre that does not have a readily identifiable name: it is a form that celebrates past genres of public oration and storytelling all the while admitting their limitations. In other words, it is neither pure appreciation through appropriation nor satire through mockery/condemnation.[9] It is a hybrid that constantly emphasizes the antitheatricality of its artifice while exalting in the pleasures it affords as if against its will. The printer, Richard Jones, included a note in the 1590 Octavo and Quarto editions that seems to affirm this point. And it does this, as it, perhaps unwittingly, contributes to Marlowe's deceit conceit of *Tamburlaine* as post-theater:

I have purposely omitted and left out some fond and frivolous gestures, digressing and, in my poor opinion, far unmeet for the matter, which I thought might seem more tedious unto the wise than any way else to be regarded, though haply they have been of some vain conceited fondlings greatly gaped at, what times they were showed upon the stage in their graced deformities. Nevertheless, now, to be mixtured in print with such matter of worth, it would prove a great disgrace to so honorable and stately a history. (189)

Although Jones makes the point that these comic scenes, which he has excised, are more suited for the mixed audience of the theater, he is really addressing the mixed nature of the play's genre in general. It is a genre that admits the ridiculousness of an earlier form of theater through the use of comedy, the "frivolous gestures," that seem to be "digressing" from the tale at hand. When the tale at hand is *Tamburlaine*, however, one cannot excise all of the digressions because the tale itself celebrates and is precisely what Jones labels as "graced deformities." Far from creating a play that can be split into an "honorable and stately ... history" and that which "vain conceited fondlings greatly gaped at," Marlowe's *Tamburlaine* in fact rejoices in the antitheatricality of staging the "graced deformities" themselves.

The case of *Tamburlaine* proves Marlowe to be a transversal thinker and writer by co-opting the antitheatricalist rant and making great fugitive theater out of it. Yet, Marlowe's appropriation is not an adoption without revision; he rewrites the antitheatricalists' claims in such a way that dismantles, reconfigures, and expands on them. Unlike the antitheatrical arguments made by men like Phillip Stubbes and William Rankins, which contend that the theater is a dangerous site because it can transform and move the senses to ill effect, Marlowe's fugitive antitheatricalism in *Tamburlaine* maintains that neither seeing a spectacle nor hearing a wonderful piece of oration can guarantee transformation. *Tamburlaine* celebrates the "mirror" that can never fully merge with the reality that it reflects, and posits a powerless theater because it can always be said to signify nothing, like the tale told by an idiot in Shakespeare's *Macbeth*. The fact that this virtually actionless antitheatrical performance was so popular on the Elizabethan stage may be evidence for the transversal power of Marlowe's post-theater. *Tamburlaine*'s subsequent lack of a popular stage history may indeed reinscribe Marlowe's own fugitive exploration. Whereas *Tamburlaine* had a long run during the sixteenth century when the antitheatricalists held sway, the seventeenth, eighteenth, nineteenth, twentieth, and twenty-first centuries have not had such strong antitheatrical forces and *Tamburlaine* has subsequently faded offstage, if only to reemerge significantly in the Marlowespace that has come to occupy much academic discourse today.

Notes

1. To give some examples: Jonas Barish argues that "Tamburlaine ... makes the battlefield into a stage, with captive kings and royal potentates as audience." Jonas Barish, *The Antitheatrical Prejudice* (Berkeley: University of California Press, 1981) 186. Johannes Birringer writes about "the immense theatricality of the play." Johannes Birringer, "Marlowe's Violent Stage: 'Mirrors' of Honor in *Tamburlaine,*" *ELH* 51 (1984): 227. David Fuller addresses Tamburlaine's "self-fashioning ... Putting on his costume in full view of the audience." David Fuller, "*Tamburlaine the Great* in Performance," *Marlowe's Empery Expanding his Critical Contexts*, eds. Sara Munson Deats and Robert A. Logan (Newark: University of Delaware Press, 2002) 67. Marjorie Garber argues that the audience hears "the dangerous doubleness implicit in [Tamburlaine's] words, the metatheatrical or metapoetic referent." Marjorie Garber, " 'Here's Nothing Writ': Scribe, Script, and Circumscription in

Marlowe's Plays," *Theatre Journal* 36 (1984): 303. And finally, Stephen Greenblatt has suggested that Tamburlaine, like many of Marlowe's protagonists, "repeats himself in order to continue to be that same character on the stage. Identity is a theatrical invention that must be reiterated if it is to endure." Stephen Greenblatt, *Renaissance Self-Fashioning From More to Shakespeare* (Chicago: University of Chicago Press, 1980) 201.

2. Christopher Marlowe, *Tamburlaine the Great, Part I* (Prologue lines 1–8), *English Renaissance Drama: a Norton Anthology*, ed. David Bevington, et al. (New York: W. W. Norton & Co., 2002). All citations from this play will come from this edition.

3. Johannes Birringer, for example, argues that Marlowe is attempting to set himself apart from the "rhyming motherwits." He writes, "After only a few lines the Prologue has made it clear that this play will initiate a radical break with the theatrical fashion of the day ... The self-conscious annunciation of an agonistic principle of Word and Sword, and especially the exulting manner in which the new poetry is introduced, sets up a new horizon of expectation for the audience ..." (222).

4. Greenblatt 202.

5. Jonathan Crewe has an interesting article about the connections between Marlowe's *Tamburlaine* and William Rankins' antitheatrical pamphlet, *A Mirror of Monsters*. Jonathan Crewe, "The Theatre of the Idols: Marlowe, Rankins, and Theatrical Images," *Theatre Journal* 36 (1984): 321–33.

6. The original lines are: "Was this the face that launched a thousand ships / And burnt the topless towers of Ilium?" Christopher Marlowe, *Doctor Faustus* (5.1.90–1). *English Renaissance Drama: a Norton Anthology*, ed. David Bevington, et al. (New York: W. W. Norton & Co., 2002).

7. David Fuller disagrees with our assessment of the lack of action occurring onstage. He poses that the stage directions about the battles are ambiguous and suggests that "an audience of *Tamburlaine* that never sees any actual fighting is likely to feel that the restraints of classicism are being wildly misapplied" (70–1). We feel, however, that the "restraints" that Fuller describes are part of the strange generic tension that the play works so hard to establish.

8. Marjorie Garber, for example, writes that while Tamburlaine sets out to "rewrite history," he finds "himself at the last revised, re-signed and rewritten by the unseen rival he cannot overreach, the author who makes the prologues and epilogues" (308). In a similar vein, Richard Schoch contends that "Tamburlaine's predicament in language – he is the oracle whose authority is only ever second hand; he ventriloquizes working words which already belong to someone else – enacts the tension between representation of action in the dramatic text and the actuality of action in theatrical performance ..." Richard Schoch, "*Tamburlaine* and the Control of Performative Playing," *Essays in Theatre* 17 (1998): 6.

9. On this topic, we are indebted to Jonathan Gil Harris, who, in conversation with Ayanna, likened *Tamburlaine* to Quentin Tarantino's *Kill Bill*, a film that admits the limitations of its genre (the kung fu movie) while nonetheless celebrates it and presents it earnestly.

Works cited

Barish, Jonas. *The Antitheatrical Prejudice.* Berkeley: University of California Press, 1981.

Birringer, Johannes. "Marlowe's Violent Stage: 'Mirrors' of Honor in *Tamburlaine.*" *ELH* 51 (1984): 219–39.

Fuller, David. "*Tamburlaine the Great* in Performance." *Marlowe's Empery Expanding his Critical Contexts.* Eds. Sara Munson Deats and Robert A. Logan. Newark: University of Delaware Press, 2002. 61–81.

Garber, Marjorie. " 'Here's Nothing Writ': Scribe, Script, and Circumscription in Marlowe's Plays," *Theatre Journal* 36 (1984): 301–20.

Greenblatt, Stephen. *Renaissance Self-Fashioning From More to Shakespeare.* Chicago: University of Chicago Press, 1980.

Marlowe, Christopher. *Tamburlaine the Great, Part I. English Renaissance Drama: a Norton Anthology.* Eds. Katherine Eisaman Maus and David Bevington. New York: Norton, 2002. 183–244.

Reynolds, Bryan. "The Devil's House, 'Or Worse': Transversal Power and Antitheatrical Discourse in Early Modern England," *Theatre Journal* 49.2 (1997): 143–67.

—— *Becoming Criminal: Transversal Performance and Cultural Dissidence in Early Modern England.* Baltimore: Johns Hopkins University Press, 2002.

8

Becomings Roman/Comings-to-be Villain: Pressurized Belongings and the Coding of Ethnicity, Religion, and Nationality in Peele & Shakespeare's *Titus Andronicus*

Glenn Odom & Bryan Reynolds

In attempting to comprehend associations between the character Aaron's villainy and blackness in George Peele and Shakespeare's *Titus Andronicus* (1592), we begrudgingly use the term "race," finding ourselves in what has become an all too common quagmire between impulses to exclude and include the word in discourse on social identity.[1] Recognizing race's power as a social construct with erroneous foundations in biological difference, we use the term with this qualification to address and undermine its conventional usage and application to Elizabethan theater. Perceptions of race in the period depended on physical, social, political, and religious factors, including, but not limited to, skin color, creed, gender, and nationality. Aaron, one of the first villains of the revenge tragedy genre, a status that further complicates his identification, is "black," with all the connotations that this word carries for Elizabethans; his status as a Moor bars him from the official culture of both Shakespeare's England and *Titus'* Rome. Peele and Shakespeare combine "Moor" with "black," "villain," "non-Roman," and "non-English," thereby producing a "translucent effect" in which these categories function as societal, experiential, and hermeneutic limitations; but also as markers, roadmaps, and vehicles by which Aaron and other characters in *Titus* navigate the sociopolitical terrain.[2] In their varied responses to Aaron, all of the major characters of *Titus* move across at least one kind of boundary, whether conceptual, subjective, social, cultural, or political, even while they are constrained by others. In moving across boundaries, the characters achieve new forms of empowerment and powerlessness within the play.

"Fugitive explorations" are a special kind of transversal analysis with the aim of achieving agency where agency had been wanting, inhibited, evacuated, or forbidden (see Chapter 1). It is analysis committed to understanding "fugitive elements," which are codes, themes, identities, narratives, humans, and so forth that exist in defiance of, but are still subject to, official territories encompassing, within, or connected to the subject matters examined or experienced. By virtue of being renegade, fugitive elements, especially when in the form of chemicals, people, ideas, emotions, characters, textual moments, slips in speech, and political commentaries, have potential to unleash or produce "transversal power" – any force that generates changes in conceptuality, emotionality, and physicality that undermines the organization of individuals, groups, societies, and systems – when they engage with subjective and official territories, which are the conceptual-emotional ranges from within which, respectively, a person or group typically relates to the world.[3] In these cases, transversal power's fugitivity challenges and possibly transforms aspects – such as philosophical and moral tenets – of subjective and official territories that precipitate the driving out of fugitive elements even while it simultaneously empowers them. We are particularly interested in fugitive elements within and connected to *Titus* that may have escaped the purview of other Shakespeare scholars that can work to defy their readings and empower symbolically, semantically, materially, or discursively the elements themselves. When marking *Titus'* position with respect to ethnic, religious, and national categories, the fugitive elements operative in Peele and Shakespeare's text come into focus.

Translucency in coding of identity: the category of the black man

Peele and Shakespeare tendentiously remind the audience that Aaron is black as well as Moorish, foreign, and a villain. Early modern literature frequently associates the concept of black with the bestial, sexual, and physically grotesque. In their essay on *Othello*, Joseph Fitzpatrick and Bryan Reynolds refer to a body of literary criticism that deals with or participates in these theatrical, literary, or scientific manifestations of racism, as exemplified by Allan Bloom and Harry Jaffa on the one hand and A.W. Schlegel on the other. Peele and Shakespeare display a similar negative racial portrayal in lines of Aaron's such as, "Oh how this villainy / Doth fat me with the very thought of it! / Let fools do good,

and fair men call for grace, / Aaron will have his soul black like his face"
(3.1.202–4). These lines suggest that Aaron's machinations stem from
his dark skin, and his dark skin marks him as barbarous. The concept of
blackness, however, is not a unified field. Fitzpatrick and Reynolds
explicate the characters of Othello and Iago in terms of both barbaric
and Venetian ideologies. They demonstrate that Othello and Iago move
among the subjective territories created by these ideologies, and that
the binary oppositions constructed through employment of what transversal poetics refers to as the "dissective-cohesive mode of analysis"
(d.c. mode) – which separates variables of the subject matter under
investigation in the interest of producing a totalized account of their
relationships to each other – has led to misrepresentations of the play,
as we find in much of the critical commentary on *Titus* with regard to
Aaron's blackness. Fitzpatrick and Reynolds do not interrogate the
precise content of the ideology of barbarism, and thus do not comment
on some of the other common oversights contained in the body of
Othello scholarship. They seem unaware that the portrayal of the black
man on stage, much like the portrayal of the revenger in revenge
tragedy, had not been established when Peele and Shakespeare wrote
Titus and was still developing during *Othello*'s heyday.

Elizabethan drama staged characters dressed in black, often with
black makeup. These characters were not explicit representations of
black men, but rather depictions of demons and vice. Although Elizabethans did have negative associations with black people and with the
color black, when Peele and Shakespeare wrote *Titus* they were without
a history of depictions of black men on the public stage with which to
shape Aaron. Starting in 1510, there were a series of ten court masques
(the texts of which do not survive) which depict Moors (although not
necessarily black Moors) on stage. While these masques may certainly
have inflected the representation of Moors in the public theater,
masques and pageants are a different medium, and at the very least
would require extensive adaptation to transfer into Shakespeare's
theater. Focusing then on publicly presented plays, rather than court
entertainments, only George Peele's *The Battle of Alcazar* (1588) precedes *Titus* and contains a black character. If one includes Moors and
Jews as "black," an extensively debated subject, then Christopher
Marlowe's *Tamburlaine* and *The Jew of Malta* also precede *Titus*. In any
case, few portrayals of blackness came before Aaron.[4] Peele and
Shakespeare's Aaron was not part of a theatrical tradition of portrayals
of black men. Rather, he took part in the creation of a tradition.

In terms of non-theatrical exposure to black men, Elliot Tokson states that the majority of the British had not seen an African before the 1550s when five Guineans were brought back to London (1). Discussions of blacks were included in the travelogues of Sir John Mandeville (1356) and in Sebastian Muenster's edition on geography (translated 1553), as well as in a handful of other travelogues published around the time Peele and Shakespeare wrote *Titus*. These travelogues are by no means a unified body of literature with regard to their representation of blackness. Therefore, when critics like Ayanna Thompson state such things as, "Shakespeare is one of the only revenge tragedians to capitalize on the potential connections between the corporeal and the early modern fascination with racial difference" (326), they ignore the fact that the fascination, if it can be termed such, was nascent and still being formed in theatrical representation.

A similar difficulty can be seen in Arthur Little's discussion of race in early modern literature. He argues that the semiotic associations among rape, race, sacrifice, and violence prevalent in early modern England can be seen in "the narrative and theatrical spaces and entanglements it [race] shares with rape and sacrifice" (1). He does not clarify that Elizabethan literature helped foster this entanglement; rather, Little writes of the entanglement as a conclusion from which Elizabethan literature works. Little uses what he surmises are the related terms of race, sacrifice, and rape to explain the sociopolitics of *Titus*, which he sees as revolving around the concept of "purity," as in Shakespeare's *Lucrece*. According to Little, sacrifices enable a conversion from evil to purity, or from black to white. Continuing with these binary categories, Little problematically states: "Early modern dramatic rape narratives frequently and predominately feature black men" (5). Little employs examples dating as late as the 1700s to explain the attitude toward race in *Titus*, and then relies, in part, on characters that are metaphorically black to flesh out his argument. The Senecan revival also ensured numerous rape stories that did not involve black men. Little says, "interracialness works in such narratives as a device, an ideology, rather than an actual occurrence" (5). Little argues that due to this ideological construction, while one may cross the boundary between semiotic associations of blackness and whiteness, one cannot maintain a space between them. The crossing from white to black or from black to white is, by this standard, an absolute. It does not allow room for the literal figure of Aaron's child as mixed and interracial.

According to Little, sacrifice serves to overcome a crisis of this boundary crossing; when stability, as defined by a series of dichotomies, is threatened, a pure sacrifice (Lavinia/Lucrece in this case), representing a pole of a binary, must be made. The appropriate sacrifice will act to strengthen these boundaries and reinforce the binaries. Little cites the opening debate between Bassianus and Saturninus as a contention that agrees only on the honor and purity of Rome. Both accuse the other of impurity and thus create a dichotomy. However, Marcus steps in and suggests a third alternative. When Marcus suggests that Titus rule Rome, he is not defending a binary system of identity; instead, he is finding consensus between two opposing sides. When Titus turns down the crown, the Roman state does in fact destabilize, but the binaries that create the parallel between Lavinia's rape and sacrifice are no longer present.

Although Little is correct in identifying an entanglement, the linkage between sacrifice, race, and rape is not absolute, and thus his sacrifice/conversion model of crossing dichotomous boundaries does not sufficiently explain the entanglement. Titus was chosen emperor of Rome, his family is buried in an honored vault, and he has just conquered the barbarians. If purity is equivalent to whiteness, then Titus himself, free of the touch of any black men or women, should be pure. In fact, Titus ends up killing three of his own children and three of Tamora's sons, actions which are not viewed as pure within the play. In addition, Lavinia's sacrifice comes after the Goths have captured Aaron, and therefore moments before the death of Tamora, Titus, and Saturninus (and the revelation of the deaths of Chiron and Demetrius). Lavinia's sacrifice cannot be isolated as the one that restores order.

Focusing on conversion rather than sacrifice, Ania Loomba, like Little, marks females as the site of cultural change: "We have the recurrent spectacle of a fair maid of an alien faith and ethnicity romanced by a European, married to him, and converted to Christianity. Her story, unlike those of converted men, does not usually end in tragedy, nor does it focus on the tensions of cultural crossings. ... Instead of a self-fashioning, hers is a refashioning by her Christian husband" (212–13). All women are subject to this conversion narrative, according to Loomba, particularly the queen: "The changeability, 'turnability' of the queen is thus central to her symbolic role ... the significance of the female, especially queenly, conversion or assimilation is not limited to religious otherness. The exchange of women has always signaled the vulnerability of cultural borders" (218). Little likewise claims: "Given that women, especially black women, are already seen as having a dangerously mutable racial identity, it should also not be surprising that there exists a popular tradition in early modem England of converting black-women and blackened women to white. It should also not be surprising given the black woman's tenuous hold on a 'real' identity – that is, an identity other than a radical one. Sharing a race with black men and a gender with white *women*, she always lacks a real signifying (natural and dominant) identity" (163). Consequently, while Loomba and Little agree that women mark the fear of border crossing, Little maintains that women are sacrificed to reinforce the border, while Loomba maintains that exchange of women allows border crossings to take place; she assumes a permeability of a racial boundary, yet an inflexibility of gender.

After marrying Saturninus and undergoing her supposed conversion, something in Tamora's character prevents her from being fully changed; her femaleness does not make her mutable, as Loomba argues. Lavinia believes that Tamora converts to a Roman ideology, being ruled by her new husband; and then, when it is revealed that Tamora is having a sexual relationship with Aaron, Lavinia condemns her for both infelicity and lack of womanly virtues: "O, Tamora, thou bearest a woman's face" (2.3.136). Had Peele and Shakespeare written *Titus* later, it could be argued that he subverts the potential of the sacrifice/conversion model for dramatic ends. The dramatic action of the text explores the ability of all the characters to assimilate or reject certain codes to which they have been assigned. Tamora should be able to be converted from black to white with relative ease, according to this argument. Instead, she corrupts Saturninus and converts herself, according to Lavinia, into an evil man. Essentially, the sacrifice/ conversion model, while pointing out the need to view racial difference in a semiotic field, relies on consistently related dichotomies to function, and therefore cannot account for the complexities of the script.

The sacrifice/conversion model can account for fears about racial crossing, but it does not deal with the fear or actualization of hybridity. Fear of miscegenation was present in Elizabethan England (although, as will be noted later, several of the Roman rulers depicted in plays had consorts of other races). One of the first black men to arrive in England married a white woman in the mid 1550s, and texts refer to their offspring as "a disparagement of the white race" due to his "coal-black" skin (Tokson 1). Ravenscroft's 1687 version of *Titus Andronicus* has Tamora kill her own biracial infant. In this version of the play, Aaron threatens to eat his own murdered child. Ayanna Thompson argues that the consumption of the child is a protection against the horror of the biracial corpse. It was not only false conversion that had to be warded off, but also the hybrid being who existed as part of two races (332).

It was hard work to earn the title of villain in a revenge tragedy, and these same villains also steal, lie, murder, betray, and do most every other thing considered despicable by Elizabethans. To give miscegenation precedence over the others, due to its racial nature, would require perpetration of miscegenation by the majority of the villains. Not all Elizabethan villains were black, nor were they all implicated in either rape or sacrifice. Therefore, to claim that miscegenation is the primary force operating in *Titus* is to ignore the various codes that Aaron enacts and the sophistication of his devout effort at villainy. Little and Loomba both correctly identify race as entangled with other concepts, but these entanglements are not consistent linkages; nor do categories which fail to take into account hybridity fully explain the sociohistorical context. In early modern England, to be black was not to be a villainous Moor; to be a villain was not to be a black Moor; to be a Moor was not to be a black villain; to be bestial and barbarous was not to be a black Moor. Such reductionism is characteristic of the d.c. mode of analysis.

Translucency in coding of identity: the category of the villain

The modern revenge tragedy has two antecedents: Senecan drama and liturgical and ecclesiastical dramas popular in the first half of the 1500s. But it is not merely a fused imitation of both these forms.[5] Bernard Spivack refers to the villains of the revenge tragedies as hybrids:

> a stock dramatic figure persisted on the stage during a period of rapid change away from the theatrical convention that gave him birth and supplied his existence with meaning. He himself is gradually transformed, but because of his powerful hold on the popular stage in his original form and nature, it is a compelled and reluctant transformation, lagging behind the rapidly evolving naturalism of the English drama after 1550. (33)

Depending on the specific dates of the plays, either *Titus* or Thomas Kyd's *The Spanish Tragedy* is the first of the modern revenge tragedies.[6] It is by overlooking this fact that much of the scholarship on *Titus* makes its first mistake with regard to Aaron. There is no "traditional" manner in which Machiavellian villains, and particularly black villains, were depicted on the English stage before *Titus*.

Peele and Shakespeare were not working within a tradition of revenge tragedies which they attempted to unseat with the racially marked Aaron. As noted, though, the revenge tragedies themselves do follow a period of religious drama. Prior to the secular Senecan imitations, such as by Marlowe, Kyd, Buchanan, Baldwin, Norton and Sackville, and Gascoigne and Kinwellmersh, the religious dramas presented vice as a physical character on stage. The morality plays are largely allegorical in nature, while the Elizabethan revenge tragedies are historical, in the sense that they contain plots centered on events that take place in a physical world. As Spivack shows, the hybrid villain of the revenge tragedy maintains a bit of the allegorical character of the morality plays. Devils and demons in liturgical drama were frequently presented as black: this black, however, is a blackness of costume and face paint. These characters are not representative of Moors or Africans, although certainly the linkage is nascent in form. The allegorical figure of Vice stands between the blackface devil and the revenge tragedy villain: "[Vice] becomes the familiar theatrical label for the stock role of the homiletic artist, who, as a protagonist for the forces of evil, created and sustained the intrigue of almost every morality play. The Vice is at once the allegorical aggressor, the homiletic preacher, and the humorist of the morality plays" (Spivack 135). Spivack identifies Aaron as a hybrid rogue, existing partly between the symbolic blackness of a devil and physical blackness. Aaron is neither an allegorical figure, a representation of Vice, nor an embodiment of the Devil. His black skin, then, would not have been an immediate marker of evil for the Elizabethan audiences, although they could retroactively appreciate the new symbolic connection.

Thus, Aaron's place as a liminal figure, as a villain and outsider prevents him from being precisely categorized, and thereby gives Aaron, the villain/vice, the transversal power often found in fugitivity. Aaron provides humor and acts as the vehicle of moral commentary within the play. He identifies evil and lack of virtue with much greater accuracy than the Andronici. Insofar as Elizabethan audiences did not know what to expect of this new hybrid figure, Aaron's fugitivity is identifiable within both *Titus'* Rome and Elizabethan England. The Elizabethan aspects of this fugitivity become clearer when Aaron is viewed in the context of the Moorish code, rather than that of the villain.

Whereas the purpose of liturgical drama is to demonstrate the dangers of evil and the manner in which good triumphed, revenge tragedy is fueled by notions of dissimulation and betrayal. For there to be betrayal, there must be trust. Assuming that the heroes of revenge tragedies are somewhat intelligent, they would not place their trust in a devil clearly labeled as such. Hence, as Anthony Gerard Barthelemy explains, "in order to dissemble and deceive, the vice must look like a man": the Machiavellian villain replaces the vice figure (73). As the stage became increasingly more mimetic and less religiously affiliated, this demand for dissimulation grew. While the villains of revenge tragedies no longer went by names like Sloth and Avarice, the playwrights still marked them clearly as villains: "The vice had no symbol of office to hand down to the villain, so the villain received instead the vice's characteristic dissembling and ranting" (75). Aaron's expressions of his own "evil" fall into this category of rant or monologue celebrating evil. He delights in his villainy and explains his plots in every speech he makes; the following lines are exemplary: "I am no baby, I, that with base prayers / I should repent the evils I have done, / ten thousand worse than ever yet I did / would I perform if I might have my will" (5.3.185–8). Whereas Aaron also emphasizes his blackness, the newly forming character type was not always black. Barabas in *The Jew of Malta* (1589) is Machiavellian in his deceits (according to Freeman, he is not properly a villain in a revenge tragedy [17]). In *The Spanish Tragedy* (~1590), Villuppo, the villain of the subplot, is of "Portingale," and Lorenzo, the central villain, is Spanish.[7] These villains soliloquize like the vice figures, despite their non-blackness.[8] In addition, Titus refers to Chiron and Demetrius as Rapine and Murther, thereby characterizing them as the vice figures in *Titus* (5.2.83).

The frequency and vituperative nature of Aaron's rants, nevertheless, exceeds that of Barabas, Villuppo, Revenge, and Lorenzo. Scholars have also commented on Aaron's lack of motivation for a majority of his deeds: Aaron claims he does them for a delight in evil. In Peele and Shakespeare's source text, it is Tamora, the queen of the Goths, who is the central villain. By making Aaron black, Peele and Shakespeare lessened the distance between villain and the vice figure of liturgical drama, with whom Aaron's rants have more in common than with the ranting of villains in the later revenge tragedies. Unlike vice figures, Aaron creates elaborate plots to work his will, which, like the vice figures, is evil for its own sake. This connection between villain and vice provides one reason for Aaron's blackness, and it places some of the focus on the evil deeds of Aaron rather than on his "racialized" body. This division between the black of the vice figure and black as a racial marker is not categorical; as we have seen, Aaron's status as a transitional figure between vice and villain does not explain his personal focus on black as an identity marker.

Translucency in coding of identity: the category of the foreigner

Marked by a skin-color and a character type, Aaron is also a foreigner. In her argument about the connection between race and religion, Loomba aligns the nationalities, English and Spanish, with religion, Protestant and Catholic. Jeffrey Knapp provides evidence of a different sort of pseudo-national identity at work: "Robert Parsons (1599) ... spoke of a time when 'one God throughout this little Island of England, and Scotland, but also the whole body of Christendom, one faith one belief ... but now in these point, ... we English ... are not only different & divided from the general body of Catholics in Christendom (with whom we were united before) but also among ourselves and with other new sectaries sprung up with us or after us, we have implacable wars and are divided in opinions' " (83–4). The concern here with both religion and nationality in such a short space of time suggests that in early modern England they were seen as closely connected but also distinguishable entities. This passage also shows a confusion of identity: the "we" is English, but Parsons cannot determine what it does or should mean. The need to establish an English identity, as Willy Maley illustrates, has always transcended religious difference; he identifies the "British Problem" of identifying England's culture with British culture, which challenges the imagined unity of Scotland, Britain, Ireland, and Wales (and sometimes France). Maley states that Shakespeare "lived in a polity that consisted of England, Wales, and – contested – Ireland. The royal house was of Welsh provenance, and the Irish wars were the most pressing contemporary political conflict" (86). According to Maley, Britain chose Rome as a manifestation of colonial relations and the British ideal.[9] Within the first scene of *Titus*, characters refer to Rome, Romans, the Roman emperor, and/or the Roman people once every five lines. For instance, "Romans, friend, followers, favorers of my right, / If ever Bassianus, Caeser's son, / were gracious in the eyes of royal Rome" (1.1.9–11), and, "Know that the people of Rome, for whom we stand / a special party, have by common voice, / in election for the Roman empery, / Chosen Andronicus, surnamed Pius / for many good and great deserts to Rome" (1.1.20–4). Statements about honor, justice, glory, pride, and duty tie these references to Rome together. If Peele and Shakespeare marked Aaron (and Tamora) as "other" in this play, then it is as "other" to Rome/England, and not just to a discrete racial category, religion, or concept of virtue.

What are the values of this Rome/England from which Aaron is excluded? Spivack delineates a shift between secular and spiritual values in the late 1500s (62). The values of "Rome" correspond to the Elizabethan ideas of secular virtue.[10] In this case, Elizabethans arrive at a construction of Rome via Machiavelli, Livy, Tacitus, and Plutarch. Amyot had translated both Plutarch's *Parallel Lives* and most of *Moralia* by the time Peele and Shakespeare wrote *Titus*. In these highly didactic texts, Plutarch stresses the value of moral education and public spirit-edness as well as the costs and benefits of war. Machiavelli, however, further filtered the idea of Roman virtue into his construction of *virtù*. In his study of the relationship between Shakespeare and Machiavelli, John Roe defines this as "the source of ambition and therefore, poten-tially, of vice" but also as "magnanimity" (xii). If one applies this Machiavellian logic directly to *Titus*, then the very virtues of Rome are also the cause of Titus' downfall. He is too courageous, too full of *virtù*, and like Machiavelli's Caesar, he must fall. Aaron, conversely, lacks this *virtù*. Aaron is not virtuous because he is not Roman: he is a foreigner, and therefore the virtue of Rome is foreign to him. Nevertheless, Tamora is motivated by the same desire for revenge, the same cour-ageous demonstration of love for family that prompts Titus to slaugh-ter her children. Saturninus, on the other hand, once he becomes emperor of Rome, does not take any significant action in the text: he lacks *virtù*.

The play's opening quarrel between Bassianus and Saturninus embodies the instability of Roman identity that becomes pervasive. The factions divide the empire with their feud, invoking the early modern division of what came to be known as the "Anglo-Saxon Union" (Maley 104). Maley states that the play reflects that the attempted "internal colonization" of England had to be deemphasized in support of the "British Myth" of unity and empire (103). In other words, if Britain was to emerge as an international power, Britain first had to define itself in terms of the competing interests of England, Scotland, Wales, and Ireland. The focus on an external threat, Goths and Moors for *Titus'* Rome or French and Catholics for England, allows the monarch to cover up the fissures within the state. England was officially at war with Spain during the early part of Elizabeth's reign, given that Pope Pius V declared Elizabeth a heretic. Andrew Hadfield explains that Elizabeth moved toward British unity despite the fact that Mary Queen of Scots "was associated with aggressive Catholicism and the murder of Protestants, an alliance between England's traditional enemies, underhand dealing and connivance, and civil war, in both Scotland and France" (161). With an external threat, Britain, like Rome in the play, needed to draw itself closer together, yet Elizabeth was not able to establish a unified British identity. The English continued to perceive the Scottish as barbarian threats (Hadfield 174). Rome's instability in *Titus* reflects England's at the time.

G. K. Hunter traces this change in attitude toward the foreigner to a different cause and locates the change during the same period of twenty years. For Hunter, the profusion of scientific knowledge of the world outside of England that was "brought to the national culture had to be fitted, as best it could, into a received image of what was important" (41). Hunter argues that the divine connotations of the *mappa mundi* governed the reception of the new scientific data. Just as Spivack argues that Aaron is a hybrid of spiritual/vice and secular/villain, Hunter asserts that the foreigner in Elizabethan England straddles the line between a group separated by spirituality and one separated by geography and culture. In either case, the foreigner "comes into the literary focus caught between the xenophobic poles of Fear and Derision, which had always operated where Englishmen and Foreigners came into contact, but which was new as a literary image" (46). With this argument, Hunter provides one of the clearest rationales for keeping separate the categories created by racism and xenophobia.

Translucency in coding of identity: the category of the Moor

To say that Aaron is a foreigner to Rome/England also implies that he is a stranger to the religion of Rome/England as well. The villain of the revenge tragedy adopted features stereotypically antithetical to dominant Elizabethan conceptions of goodness and purity, and these features reflected changes in cultural values throughout the early modern period. Attitudes towards Islam and Africa in general underwent a transformation in the late 1500s, and it is during this time that the villain became marked as Moorish, although not necessarily a black Moor.[11] *Titus* reflects this change.[12] In 1578, Sultan Abdul Malik, then head of the Saadian dynasty, defeated King Sebastian of Portugal at the battle of Ksar Kbir. This battle had already been staged in Peele's *The Battle of Alcazar* (1588) when he and Shakespeare wrote *Titus*, and such presentations, characterizing the Moor as a villain, were popular.[13] Despite this association, as already noted, not all villains who appeared on stage were Moorish, and not all Moors were villains.[14] In *The Battle of Alcazar,* for instance, there is a noble Moorish character; the evil Moor is black and the "dignified" Moor is white. This division between noble and savage Moor is not always along the color line as seen in Richard Hakluyt's *Principal Navigations* (1589). As Emily C. Bartels observes, "the term Moor was [used] to designate a figure ... who was either black or Moslem, neither, or both. To complicate the vision further, the Moor was characterized alternately and sometimes simultaneously in contradictory extremes, as noble or monstrous, civil or savage" (434).

We do not want to collapse the various ways in which Moors were portrayed. Rather, following the investigative-expansive mode of analysis, we want to explore the contradictions noted by Bartels and some of the different interpretive possibilities to which they point. Whereas stage presentations tended to place the Moor at extremes, either eroticizing, demonizing, or celebrating the foreigner, this was not the only exposure to Moors experienced by Elizabethans. In "A Portrait of a Moor," Bernard Harris explains that England established formal diplomatic and trade relations with Moors in 1551, while it still did not accept "black" men (24).[15] The interactions with Moors were "a compromise, in a sense of compromising. The questionable alliance was put in terms of the advice offered Elizabeth in 1586, that 'Her Majesty in using the King of Fez, doth not arm a barbarian against a Christian, but a barbarian against a heretic'" (Harris 26). The advice in question characterizes Moors as barbarian, rather than the religious label of heretic. Harris cites letters that place barbarous Moors in the houses of nobility for several months at a time. When compared to Spain, England's attitude toward Moors was charitable.[16]

In English history, images of Moors ranged from honored guest, to shrewd merchant, to barbarian, to murderer, to heathen. The question of the Moor on the Elizabethan stage has been the subject of countless articles, most taking as their focus a limited set of these historical categories. D'Amico provides a broad literary history of Moorishness, and concludes that Shakespeare "saw that with the kind of political expansion that characterized Renaissance Venice and ancient Rome came the problem of absorbing the outsider and the fear of being absorbed" (214). In transversal terms, this fear of being absorbed corresponds to the concept of comings-to-be. Furthermore, Shakespeare's treatment of Moors, according to D'Amico, demonstrates, "For the modern, cosmopolitan state that thrives on the exchange of goods and images with other nations and cultures, this conflict persists in the struggle between closed national identity and the need for intercourse with others" (214). Both D'Amico and Harris agree that the Moor is situated in a difficult position between the necessity of trade and some degree of xenophobia. This xenophobia is linked in popular literature, argues D'Amico, to nation building, and in history to the variety of factors we have already discussed.[17]

Finally, there is a subset of criticism surrounding the Moorish question that places literary and historical evidence side by side in an effort to determine a relationship between attitudes toward Moors and modern racism. William Engel's work traces a relationship between the Moor and Death in early modern art, speech, and literature: "This term [Moor] was voiced phonetically the same as the name given to the character of Death, *'Mors'* – and also the term for a traditional festive dance – morris. So too, the term for blackamoors in Elizabethan England, *moriens,* was the same as that used for the dying person" (2). Engel also traces visual and semantic connections between love and death, which he then uses to explain part of the eroticization of the Moor and the connection between crossdressing and Morris dancing. Engel's work closes with an apt warning: "we are left wondering whether the encounter we have with the concrete, with the material [representations of Moors], ever can be anything more than a confirmation – or refutation – of what we expected to find there" (12). The Moor's identity in art and literature is a series of fluid associations. Kim Hall also notes this fluidity of identity, which she connects to the feminine.[18] The fact that the Moor occupies such a site of fluidity suggests a bias among Elizabethans: the Moor is the unknown potential threat/potential ally. To force Elizabethans retroactively to make a choice between the identities that exist within "Moor" is to ignore history, semantics, and literary examples.

In the following pages, we combine close textual analysis, historical evidence, and literary history with sociopolitical theory in our examination of the ambiguous category "Moor" in Elizabethan culture and how it relates to Peele and Shakespeare's *Titus.*[19] We accept the premise that the category demonstrates the fact that no one term can contain either the historical reality or the literary representation of the Moor.[20] Historical data provides a semiotic field and a series of categories in which Peele and Shakespeare's play exists; it suggests a grounding for the political work the text may have done; and forms part of the subjective and official registers of Shakespeare, what Hedrick and Reynolds call "Shakespace," a term that describes the multidimensional "articulatory space" through which critical and popular discourses about Shakespeare interface, are imbued respectively and relationally with meaning, and subsequent communication transpires.[21]

Pressurized belongings and boundaries of discontent: motion, not contamination

Whereas villain, Moor, British, black, Roman, and foreigner are discreet categories, an individual can contain or express all of them simultaneously, even contradictorily. As the codes overlap in Peele and Shakespeare's text, they create power structures. The action of *Titus* focuses on the ability of characters to navigate these power structures primarily via what we call "pressurized belongings," which are the related and often conflicting processes of assimilation and expulsion by which one becomes a member of an alternative group, subjective territory, or official territory at the expense of one or more of its members. The new member causes overflow or reconfiguration such that not all of the extant members can remain the same or remain at all if the system is to maintain equilibrium. This is not to say that the substance of the group necessarily drastically changes or that there is transformation of the parameters within which the group maneuvers, but rather that there is only room for so many members of certain kinds.

One metaphor for pressurized belongings is the idea of an incompressible fluid, such as water, in a container of a relatively stable size. When more water is added, some water must leave, or the container must expand. The entire container is also contained within a multitude of other containers, that is, in transversal terms, within other types of territories, each filled with, or partially filled with, fluids, which for transversal theory means concepts, thoughts, emotions, presence. The containers are of varying degrees of rigidity and flexibility, corresponding to the cohesiveness of the subjective territory of the group. The compressibility of the fluid corresponds to the ability of the individual group members' subjective territories, and of the official territory that their subjective territories together comprise, to assimilate new members. Other factors such as temperature, external pressure, and chemical additions to the fluid can influence the precise changes in the amounts of fluid inside and outside of a given container and whether the addition of fluid will change the shape of the container or cause the expulsion of other fluids. These factors correspond to sociopolitical conductors, natural catastrophes, and international events – to anything unleashing transversal power with respect to particular subjective and official territories, such as the Battle of Ksar Kbir, which, in this case, influenced England's attitude toward the assimilation of Moors. It is important to note that as elements are expelled from a territory, often becoming fugitive, the remaining occupants responsible for the expulsion, in correlation to the extent of the engagement with the expelled, are marked by the process. Such marking can be seen in *Titus* as characters expel other characters from a given system. The fluid model of pressurized belongings explains this as well. When fluid leaves a container, the rest of the fluid will respond to the changed pressure or conform to the boundaries of the container, effectively occupying at least part of the recently vacated space.

The pressurized belongings in *Titus* develop within the play's world and developed out of two historical moments both struggling for equilibrium in response to rapid change and growth. Elizabethans gained knowledge of the Roman Empire primarily via Machiavelli. An analysis of *Titus'* Rome, by extension, is an analysis of Machiavellian Rome (although, as D'Amico points out, this may demonstrate a common lineage through Livy and Plutarch rather than Shakespeare's study of Machiavelli himself). In brief, Machiavelli states that the Roman Empire is bound to collapse because the same ideals that made the empire great cannot apply to the expanded empire: contact with new cultures simultaneously strengthens the Roman Empire and harms the Roman identity. *Titus'* Rome, like Elizabethan England, engages rapidly with new groups. If the Roman/Elizabethan system is to reach equilibrium and their identity is to achieve stability, as the principle of pressurized belongings states, any new factor assimilated into their space must either be matched by an expulsion from this space or the extant space must expand in order to accommodate the addition. Expulsion will leave an empty space into which a new factor will naturally flow, while expansion changes the parameters of the system, possibly transforming it into something radically different.

Concerning identifying concepts and codes, G. W. F. Hegel argues that any notion of totalization "in such a word [as " 'the Divine,' 'the Absolute,' 'the Eternal,' etc."], even the transition to a mere proposition, contains a *becoming-other* that has to be taken back, or is a mediation" (11).[22] Following Hegel's rejection of there being anything absolute and infinite, cognitive or otherwise, Deleuze and Guattari refer to the process of "becomings-other" in contrast to stable notions of identity.[23] A subject must continually become the identity they seek in order to maintain or pass through this identity, which is to say, we would add, that the subject must convincingly perform the identity for social affirmation – for others to perceive the identity as real. It is the contention of transversal theory that this active becomings of something else, an individual's identification of codes and an attempt to emulate them, does not account for the additional codes that have aggregated around that something. The additional codes that an individual assimilates constitute a process of comings-to-be, a passive uptake of related codes. All becomings carry with them comings-to-be (see Chapter 1). This idea is also present for Shakespeare, although, alternatively, Loomba claims that it functioned in only a single direction: "Thus blackness (both as moral quality and skin color) can more readily contaminate whiteness rather than itself be washed into whiteness" (211).

Loomba's contamination model would have the "black" despoiling the "white." She argues that the foreign female can be converted to the new culture while the foreign male has the power to corrupt or rape the white female away from her own heritage: "The exchange of women has always signaled the vulnerability of cultural borders" (218). According to the historical evidence put forth by Loomba, Shakespeare's audience would have expected Tamora to convert fully to Roman doctrine, but she "resist[s] such shrinking … [her] liaison with the black Aaron problematizes her 'incorporation' into Rome" (219). Thus there is a tension between Peele and Shakespeare's portrayal of Tamora as the woman giving in to Rome and as a black figure contaminating Rome. This is a fruitful tension, as the play forces a reexamination of a simple division between male and female, and it is precisely this tension that allows Tamora to move across boundaries.

The same sorts of tensions occur in the case of Aaron. He is expected to act in villainous ways. He does practice deception, yet this deceit does not act to challenge subjective territories. Tamora's deceit is unexpected; she claims to support Saturninus, and, in Loomba's view, as a female she should be subject to conversion. As a result, when Tamora is deceptive, she is able to remove Titus from the official territory of Rome, and create a child, which challenges the stability of this territory. This is what separates simple deceit from what transversal theory refers to as a "deceit conceit" (see Chapters 3 and 4). Both may be elaborate plot, but a deceit conceit relies on preconceptions of a subjective territory in order to work its will, which in turn leads to a destabilization of the subjective territory. If Elizabethans associate the feminine with comings-to-be, then they would have expected Tamora to come-to-be all of the codes Saturninus required of her. Saturninus watches Tamora make her bid for power, which comes across as passive submission to Saturninus and Titus: "Sufficeth it not that we are brought to Rome / to beautify thy triumphs, and return / Captive to thee and to thy Roman Yoke" (109–11). Yet, in this bid, she adopts the rhetoric of Romanness used by the Romans.

Not recognizing the scope of the pressurized belongings at work, Saturninus believes that Tamora is loyal to him, that she has come-to-be Roman. Pressurized belongings can be symptomatic, accidental, or tactical. For instance, Tamora assimilates the Roman identity in order to expel Titus from it. Strategically, she compares the Goths to the Romans ("O, if to fight for king and commonweal / Were piety in thine, it is in these" (1.1.114–15]) to make a mother's plea for a son, and in doing so she further emphasizes that both she and Titus share a love for their offspring. She demonstrates her apparent assimilation by her invocation of the gods: "Wilt thou draw near the nature of the gods?" (1.1.117). But it is not Titus's gods, or the gods of Rome she speaks of; it is "the" gods, all of whom she equalizes. When Titus denies her request, Tamora charges him with "irreligious piety" rather than cruelty (1.1.130). With this accusation, Tamora asserts that only one set of gods exists, and that this pantheon is the same one that Titus worships, or should worship, if he were Roman enough.

Saturninus allows Tamora's attempt at crossing over into Roman-hood to succeed, and in doing so allows her to cross other related boundaries as well. Before Tamora becomes a Roman, he places her "like the stately Phoebe 'mongst her nymphs / dost overshine the gallant'st dames of Rome" (316), but remains acutely aware of her different skin color: "A goodly lady, trust me, of the hue / that I would choose were I to choose anew" (261–2). Despite Saturninus' powerful acknowledgment of Tamora's different hue, the difference is transcended later in the play, as one Goth states: "where the bull and cow are both-milk-white, / They never do beget a coal-black calf" (5.1.31–2). Here, Tamora is described as "milk-white." Unless the time spent as Empress has caused Tamora to lose her tan, this lends further credence to the idea that perception of color difference depends on the relative positioning of the subject to Rome. While Saturninus holds Tamora captive, her skin tone appears to Saturninus as another hue. When the child, clearly not of Roman blood, is born, Tamora suddenly appears much closer to Saturninus, "milk-white," than to Aaron, "coal-black": she has married Rome, embodied in Saturninus, and therefore must appear Roman. In becoming Roman and thereby occupying legitimate conceptual and emotional space within Rome's official territory, she has come-to-be white.

This is only a portion of what occurs: by facilitating the becomings-Roman of Tamora, Saturninus has come-to-be villainous. Saturninus, as Emperor, is the vessel of Roman identity, and by incorporating Tamora into the state, and presumably into his bed, he has allowed a decisively non-Roman element into the system. As he and the system are one, Saturninus is corrupted by Tamora. It is Saturninus, and not Tamora, who first states his desire for sexual intercourse: he says he will not leave the square until he is married and that once married, he will go immediately to "consummate our spousal rites" (1.1.338). Furthermore, Saturninus displays a bloodlust that only Tamora's council sates. In the gendered language used by Loomba, it is the masculine, not the feminine, that Peele and Shakespeare show passively moving between codes. Thus, while Loomba is correct in identifying gender as part of a matrix that affects various boundary crossings, the alignment of gender with a particular type of crossing does not hold true in *Titus*.

In this case, Saturninus expels the Andronici, particularly Lavinia, from the concept of Roman honor: "No, Titus, no, the Emperor needs her not, / Nor her, nor thee, nor any of their stock. / I'll trust by leisure him that mocks me once, / thee never, nor thy traitorous haughty sons, / confederates all thus to dishonor me" (299–301). In the expulsion, Saturninus catches the very disease he desires to expel. This expulsion simultaneously allows Tamora to assimilate the Roman code. Something must be cast out in order for the rest of that which constitutes Tamora to be affirmed within a given code. The silencing of Lavinia also demonstrates the linked nature of assimilation and expulsion. Lavinia comes-to-be barbarous, in the sense of losing her ability to speak (she can still write and read), in order that Titus may see what has happened to Saturninus. This recognition allows the expulsion of Saturninus and the enthronement of the once-expelled Lucius. Lucius assimilates back into the Roman code, but only at the cost of the sacrifice of most of the Andronici, an unfortunate consequence of pressurized belongings when sociopolitical stakes are high.

Saturninus's experience emphasizes the connection between assimilation and expulsion in the process of pressurized belongings, a process, incidentally, that when operative in politics precludes more democratic approaches. If expansion is not a possibility, the process of assimilation requires expulsion, either subjectively, or, as in this case, in regard to another character. Saturninus follows Tamora's lead and expels Titus to make room for Tamora. The process of expulsion, nevertheless, is also one of assimilation. Saturninus cannot use the charge of dishonor to expel Titus without himself becoming implicated in this charge. Tamora swears allegiance to Rome, and vows to "nurse" Saturninus as "a mother to his youth" (1.1.332). In Peele and Shakespeare's Rome, the honorable qualities are neither purity nor spirituality; they are "justice, continence, and nobility" (1.1.15) and "love and honor" (1.1.49). Hence, in order to defeat Rome, Tamora does not have to convert Saturninus in religion or in sexuality. Instead, she must defame his honor, nobility, and concept of justice. Tamora's false oath of allegiance and her relegating Saturninus to the position of child, as well as her later cuckolding of him, all work to serve these purposes in a combined deceit conceit. With honor, nobility, and justice removed from the head of state, the Romanness abandons Rome.

Despite Tamora's power over Saturninus, Peele and Shakespeare do not present Tamora as possessing characteristically masculine traits, although they do link her to the rape of Lavinia. A desire to remove Lavinia's honor motivates Tamora: "This minion stood upon her chastity, / upon her nuptial vow, her loyalty / ... And shall she carry this to her grave?" (2.3.124–6). Lavinia cannot just be killed, she must be violated, or she will not fall outside of the Roman identity. Tamora must expel the Andronici in order to protect her own Romanness. Tamora's sons, intimately linked to their tigress mother by the "milk thou suck'st" (2.3.144), take the deflowering of Lavinia further and remove from her marital chastity, as well as her hands and her tongue, with which she could serve the state. Once Tamora convinces Lavinia of her intent, Lavinia remarks, "Tis true, the raven does not hatch a lark" (2.3.149). Divested of the appearance of Roman virtue, Tamora becomes a blackbird. Lavinia contrasts the raven to the lark. Elizabethans commonly kept larks as pet songbirds. Peele and Shakespeare marked Tamora by color and, more importantly, by a resistance to conversion to Roman culture. She does not breed larks; her boys cannot be kept in the cage of Roman virtues. Her crossing is thus complete and incomplete: she becomes Roman and comes-to-be a villain.

A different alignment occurs when one examines the situation of Titus himself. Modern productions of the play stress the communal nature of Roman society (Babula 330). Characters introduce themselves via their family lineage; Marcus repeats the proud history of the Andronici, Lavinia laments the connection between Tamora and her children. For the Andronici, this familial responsibility is a role within the Roman state. Titus abdicates his responsibility as father of the state, going against the stated wishes of the Roman people: "the people of Rome, Who's friend in justice thou hast ever been / Send thee, by me, their tribune and their trust, / This palliament of white and spotless hue" (1.1.179–82). In refusing his duty, Titus refuses, literally in this case, whiteness.

In the case of Titus, the connection between positively associated codes is strong. Once he rejects Rome and the white palliament, he quickly comes-to-be a villain (in the sense of killing many people), and he is separated from the received religion of the state. Loomba cautions that such connections are not absolute in all circumstances: "Categories such as 'A Christian Jew' or 'a Turke, but a Cornish Man borne' attested to the fact that religion and nationality were affiliations which conversions could not entirely erase" (210). An active conversion will not, perforce, obfuscate prior becomings and comings-to-be. If you remove an English man from England and the proper English church, other markers of identity may fade as well, just as Titus refuses the white emblem of Rome and then loses multiple other codes. To return to the metaphor of fluids, the specific properties of a given fluid depend on the temperature, pressure, and chemical composition of the environment in which it is located. If you change this environment, the fluid will react in a variety of manners.

Titus' progression could conceivably be explained through the type of contamination to which Loomba refers. When Titus removes himself from Rome's protection, he becomes contaminated by the black evil. However, this does not account for Tamora's simultaneous change. When Titus withdraws himself from the Roman state, he provides an opening for Tamora's entrance. Tamora matches her assimilation of the code of Romanness with a further expulsion of Titus from this code. Despite Loomba's argument that black would contaminate white, Tamora is viewed, by the Goth, as having become snow-white after her coupling with Saturninus. Aaron displays his first positive characteristics in the play after touching his baby, a sign of miscegenation. In addition, the contamination model implies a passive exchange of codes. Tamora and Titus both display active and passive assimilation of codes.

A history of pressurized belongings: not actors, gypsies, women, Moors, or foreigners

In *Becoming Criminal*, Reynolds demonstrates that both majority and minority cultures in early modern England were subject to infiltration. According to contemporary sources, Elizabethan gypsies "copy the 'Egyptian rogues' in devising their own language" whereas the mainstream English population frantically made laws to prohibit the further spread of this foreignness (29). These laws are evidence of expulsions, and such laws, as in this case, generate the very discourse they attempt to limit. The gypsies are said to have assimilated what was believed to be the codes of the Egyptians, which marks them as foreigners. In expelling the English vagabonds who have chosen to become gypsies, the English come-to-be defined against these foreigners. While these comings-to-be are passive, they resulted from the active expulsion of people identified as gypsies, who were characterized as actively having assimilated Egyptian codes.

Similarly, pressurized belongings can also be seen in the case of Moors in early modern England. England chose to expel Catholicism, as embodied in Spain, which allowed greater contact with Moors. As Moors assimilated the new codes of England and became English they also came-to-be black (the Moor was portrayed artistically in English style, Moors lived in English houses, and traded for English money in English markets). These comings-to-be were gradual and in-process when Peele and Shakespeare wrote *Titus*. The contamination model is not able to account for the Moor's change in status as black. Britain's official territory determined this characterization at the time the Moors entered. The contamination model, because of its one-way orientation, also ignores the Elizabethan's fear of deceptive crossings of boundaries: not the fear that the English would passively or actively assimilate the codes of the non-English, but that outsiders could somehow pass as English/White/Christian. We have already touched on this fear in our discussion of Jews and Moors. The same pattern can also be seen in the less contentious realm of the fears of antitheatricalists and the secularization of the theater. In *Becoming Criminal*, Reynolds points out the "transcendant" power that antitheatricalists, Stubbes in this case, attributed to the stage: "Then these goodly pageant being done ... then begin they to repeat the lascivious acts and speeches they haue heard, and thereby infect their minde with wicked passions, so that in their secret conclaues they play the Sodomite, or worse" (127). This "or worse" suggests something beyond the sins represented by the vice figures and devils onstage. The boundary between fact and fiction would blur, and, by crossing this boundary, a new evil would be created, something beyond mere sin and outside of the identifiable world. If vice figures could pretend to be good, they could lead honest men astray, and yet this dissimulation of evil became the fashion on the revenge tragedy stage. Nevertheless, the antitheatricalists' fears go beyond this, as they reflect England's anxious development of a new identity:

The antitheatricalists were attempting to self-fashion themselves (in Greenblattian terms) with their polemics against the theater. They define themselves in opposition to the theater, in the names of

God, the church, and state power ("the Commonwealth") and against "something perceived as alien, strange, and hostile" (the public theater and the Devil). In doing so, they ironically experience "some loss" of a previously constructed "self." This loss is evinced by their contradictory logic, their inconsistent performance, and their argument for a fixed social identity. As it turns out, theirs is really an argument for the *fixing* of social identity that takes for granted the possibility of identity (re)construction in opposition to the identity desired by the state machinery and official culture. (Reynolds, *Becoming Criminal*, 139)

The antitheatricalists tried to form an identity through the process of expulsion and in effect became marked by the very thing that they tried to expel. This creates a homology between the antitheatricalists and the government in their relationships to theater and to foreigners.

The Elizabethan attitude toward seeming-to-be and being, in politics and on stage, was often contradictory. Dissimulation could lead to fact and to danger, yet the danger of dissimulation was precisely that it was not fact. In other words, a deceit conceit functions on a dissimulation, a seeming-to-be something other than what something appears to be, but this seeming-to-be is not, in itself, the purpose of the deceit conceit. It is not the presence of the identity behind the seeming that gives the deceit conceit its power. Rather, it is the dissimilation itself. If someone can seem-to-be part of an official territory without actually belonging to it, than the idea of the official territory is itself called into question. Early modern England's dominant ideology maintained that God orders and Satan confuses. In the case of early modern English culture, the alignments of identities created by official territories were affirmed by a concept of God and it took an alternative concept, the force of Satan, to break them down. The perpetrator of a deceit conceit, then, is doubly outside of God's plan. The very act of the deceit conceit is a dissimulation which can undermine the logic to official territory and harm specific targets within it.

The confusion surrounding dissimulation and race within the period helps to cement the racial markers to the Elizabethan concept of the villain as represented in dramatic literature. According to the Elizabethan model, the Devil is the author of vice. Vice, in its new guise as villain, acts via dissimulation. Elizabethans feared the dissimulation of the Moor and the black man. If both the Devil and the Moor are potential dissimulators, then it is an easy step to conflate the two. As noted, however, dissimilation and deceit conceit work best when least expected (the more improbable an event the greater effect it often has). This, in part, explains why all Elizabethan villains were not black or Moorish and why all Moors in dramatic literature were not villains. If dissimulation is a primary fear, and playwrights wish to capitalize on this fear, then no villainous markers should be conventional.

Oh "other" where art thou? Or why Aaron is not "other"

It would be easy yet also irresponsible to dismiss the complexity of the comings-to-be, becomings, assimilations, blendings, expansions, expulsions, and markings that delineate pressurized belongings in the period's dramatic literature as merely the product of early modern England's need to create an "other" as something against which to define itself. Movements among articulatory spaces create and are often enabled by an unpredictable force, transversal power, which is always *in potentia*, if not already in circulation. Unlike "open power," which is undifferentiated, "state power" is any force that acts to promote cohesiveness and stability in anything, whether an assortment of atoms, chemicals, words, or people. For example, in Freud's psychoanalytic theory, state power operates in his wish to bring the unbridled and unwieldy drives of the unconscious into consciousness so that the self can have meaning and order. For Lacan, state power can be seen in his theory of identity formation known as the "mirror stage," which is when a baby between six and eighteen months experiences its self as uncoordinated and fractured, but upon seeing its image in relation to another person's in a mirror imagines its self as the whole being to which it now continually but hopelessly aspires. For transversal poetics, identification, especially if empathetic, with an other – another person, thing, concept, or event – can generate transversal power and movement. This can lead to a breakdown of state power, which occurs through altering or expanding the subjective territory of the subject through an interaction with "subjunctive space," a conceptual domain in which the subject ponders "as ifs" and "what ifs" – in this case, relating to someone else's or something's subjectivity, whether real or imaginary.[24]

From within the official territory of the government, this breakdown is negative, even if the result is a new stability: official territories function on a principle of equilibrium. Assimilation and expulsion are generally matched processes. When they are not, such as when expansion is significant, the official territory itself must adapt and reform. Both Britain and Rome were undergoing this type of transformation during the historical periods in question. As Britain expanded it had to reconceptualize official and subjective territories to cope with increasing immigration and new colonial relations. The concept of other provides one possible impetus for this change. In other words, the perception of an other, an outside, provides potential inspirations and gateways for subjunctivity and transversal travel, which, given their unpredictable nature, can be at once desirable and dangerous. In this case, however, the other is not Freud's potential object of desire by which one can measure the self's adherence to its non-incestuous reproductive hetero-sexual model for a stable, "normal" sexual-gender identity. Nor is it a Lacanian transcendental "Other" (also called the "Phallus" or "Law-of-the-Father") that fills in symbolically as the center of a self-affirming, language-based system that gives only temporary stability to an individual self because the self can only ever be simultaneously other, as in "self/other," a dichotomy in which the individual continuously mis-identifies one's self with an other (as in whatever is "not-me" rather than an actual other person), and then another, forming a chain of misidentifications in an endeavor to satisfy the individual's sense of "Lack." This is a never-ending Lack that one acquires with entrance during the mirror stage into the Symbolic order of language that struc-tures all desires and is itself predicated on a concept of Lack that can never be filled. Alternatively, according to transversal theory, desire is not predicated on lack; self-formation is processual through becomings and comings-to-be and therefore always multiple (we have selves); and definition can just as easily be achieved through positive differenti-ations rather than negation, given that multiple possible "others" exist and can exist.

The "self/Other paradigm," Lacanian or in its other common modal-ities, where "Other" (usually) with a capital "O" (although sometimes with a lowercase "o" in Lacan-like applications) refers to actual others (people, things, concepts, events) and implies a consistent dichotomy as well as a binary opposition. If you arrange the categories we have been discussing in terms of binary oppositions, such as white/black, good/evil, Roman/Non-Roman, Christian/Moor (heathen), it is impor-tant to note that the alignment found in the case of Aaron of evil-black-non-Roman-Moor is not consistent across other characters. For instance, Saturninus is white and Roman but acts in an evil manner relative to the "Roman" idea of virtue. It is also important to keep in mind, as we progress with our consideration of applications of the concept of "Other" to *Titus*, that applications of psychoanalytic theory to characters in literature is necessarily problematic and are not ade-quate tests of the theory because psychoanalysis is primarily based on and intended for the study of actual humans and not fictitious ones who are the creative products of any number of humans, with any number of psychological illnesses. As the plot of *Titus* unfolds, it becomes increasingly difficult to determine which category of a hypothetical dichotomy any given character fits into. Tamora becomes Roman by virtue of marriage; she enters the official territory of Rome, but is not Roman in her heart; she never fully adopts the Roman ideals within her subjective territory. Her acts against the Andronici are, in her view, judicious revenge for her slaughtered child. If, as Maley argues, Peele and Shakespeare's *Titus* is involved in England's project of national identity-forming, then we cannot conclude that such an identity is formed through a self/Other dichotomy.

The simple binary opposition of white and black implied by the structure and content of Loomba's argument and the self/Other dichotomy, generally speaking, are often characteristics of dissective-cohesive analysis. Fitzpatrick and Reynolds explain the difficulty of this mode in terms of *Othello*: "The relationship ... is therefore reduced to symmetrical antithetism: the semblance of wholeness or totality is constructed methodologically through the binary oppositions (light/dark, white black, insider/outsider, self/Other, good/evil) either cited from or imposed onto the text to give an explanatory coherence to the characters' relationship, and to the play overall" (55). The d.c. mode's reductionism thwarts transversal poetics' investigative-expansive mode of analysis; it discourages exploration of internal connectedness and differences within a text in relation to other forces outside the immediate text in coordination with a fluid openness to reparameterizing in response to the emergence of unanticipated problems, information, and ideas. Our resistance to the Lacanian binary of self/Other shares these concerns, even though the poststructuralism of Lacanian psychoanalysis allows for more opportunities for fugitive explorations than the dissective-cohesive approach of Freudian psychoanalysis, whose primary mission is to channel the drives and desires of the unconscious into consciousness so they can be ordered and made sensible as a means by which to give stability and coherence to the self, the "I," a conscious, rational identity that supersedes the presence of the unconscious (as Freud famously exclaimed, "Where It [meaning the unconscious] was, I shall be": "*Wo Es war, soll Ich warden*").

Not dependent on either a foundational concept of Lack or negative relations per Lacan, the principle of pressurized belongings explains the structure of *Titus* and the historical treatment of race in Elizabethan England in terms of positive differences, where emphasis is on becomings and presence, even if that presence is singled out by its absence – a "present absence." By charging Titus with poor religious practice, Tamora claims to know how to serve the gods – the Roman gods – better than Titus. In effect, she expels Titus from the Roman identity, Rome's state machinery, and Rome's official culture; and Saturninus soon makes this expulsion official. Yet, in *Titus*, the process by which pressurized belongings occur marks the agents of expulsion as well. Tamora herself shows "irreligious piety" in her disregard for her infant's safety and her malice against the various Andronici. Each expulsion carries with it the necessity of a change in relation to what is expelled. A member of a group cannot be effectively displaced and replaced by another without there being affinities between them, and if the affinities are not already present, then they come-to-be, and they can come-to-be to an exaggerated degree and in unanticipated ways. The comings-to-be can cause the assimilator to adopt traits of the expelled that are unwanted and potentially self-defeating. Such situations happen antithetically to Lacan's concept of the *"object petit autre"* (aka *"object petit a"*) in his theory of identity formation, which refers to the quality projected by an individual onto an Other that works to enable the individual to properly differentiate himself or herself through negation.[25]

For Lacan, the unconscious is structured "as a language" in terms derivative of how Ferdinand de Saussure talks about the arbitrary connection between signifiers and signifieds, which form signs. Signs are the bonded entities, such as "treeness" that the spoken word "tree" and the concept/object "tree" together formulate. For Saussure, language itself is structured by the negative relations among signifiers, and therefore the signs to which they contribute. The definition of each signifier is determined by contrast with the signifiers it is not. Nevertheless, Lacan is primarily interested in relations among signifiers since, in his view, in the unconscious there are no signifieds – no stable relations between signifiers and signifieds – but rather only negations, active differentiations among signifiers (*Ecrits* 194; *The Seminar, Book III* 39–43). (Although we long to see data that supports this account of the unconscious, should we follow Lacan, the data could only ever be compensation for Lack, a fantasy projection rationalized to account for what we can never know. This is not just in this case or in the case of Lacan's own theory, but because nothing is except what it is not, something we can never define until we have contrasted it with everything else.) In Lacan's unconscious, conversations occur among four actors: the Subject (meaning the subject of the unconscious that speaks the language that constitutes the unconscious), the subject's objects, the other, and the Other's objects (such as *object petit a*) – detailed definitions of which are not relevant to the present discussion. In psychologically healthy individuals, the Lacanian Other is in communication with the Subject, even if the Subject does not know the Other. It is, in fact, a vital part of the development of the subject ("subject" with lowercase "s" as in conscious subject: the "I"/ego/self-in-process) and is in constant contact with the subject as the subject desperately and futilely imagines things that make it seem whole. For instance, in an attempt to extend the self/Other paradigm to sociohistorical inquiry as scholars often do, to say that "Africans" became the Other for "Europeans" would be to say that Europe believed it had fully incorporated Africa into its collective unconscious. Furthermore, as Lacan says in *The Seminar*, (Book III), the conscious subject is not directly responsible for speaking, but instead its speech is mediated via the internal conversations, so the therapist receives the information in "inverted form" (23).

Notwithstanding that it is not our purpose to critique misuses of Lacan's system, we want to point out that the common application of his concept of the Other in race theory, particularly in Shakespeare and early modern studies, does not take into account the complexity of his construct. The following application by Bartels is representative:

> although, if not because, the Moor was sometimes assumed to be civilized rather than savage, white or tawny rather than black, he was nonetheless circumscribed as Other. For what emerges as a key focus of "othering" within Rennaissance depictions of Moors is behavior that paradoxically ... showed them too like the English – behavior that might undermine England's claim to a natural dominance and superiority. (435)

The Lacanian Other is unknowable. The Moor was very knowable. Images of Moors and people labeled as Moors appear throughout English society. According to Lacan, "othering," what he might have called "Othering" if he subscribed to such an idea, cannot occur self-consciously. Moreover, the Other is essential to Lacan's concept of the self; the inability to receive communication from the Other is one way of defining the mental problems Lacan deals with. To say that the Other might undermine the English identity, which relied on perceived superiority, does not follow from Lacan's use of the word Other.

We have already discussed why a simple binary of self/Other does not adequately explain the social dynamics of Elizabethan England.

We have also begun to show with the example of Bartels that the uptake of the transcendental Other in race theory can be problematic. What remains to be explained is the incompatibility of pressurized belongings and Lacan's understanding of the Other. We will turn to a specific example of the application of psychoanalysis to *Titus*. Imatz Habib states that a firm perception of the self relies on the perception of the Other (for consistency, we will continue to use the uppercase for "Other," even when critics we cite do not), which can lie at the root of racism: "If a 'nation' is written always on the 'other', the human dis-possessions on which the empire is inevitably erected, the subjugated ethnic is simultaneously the desire and the danger of imperial ambi-tion and corresponding to that, the necessity and the dilemma of the master city and its domain" (91). This argument does place the self in dialogue with the Other, and, while it allows the self autonomy in its dealing with the Other, it is not un-Lacanian in its usage. Habib identifies Queen Elizabeth's attempt to deport the Other as "nervous" (93), which encourages an assessment of whether a Lacanian concept of "Other" operates in *Titus*.

Habib notes the simultaneous desire and fear of the Other, creating a dilemma for the colonial power. In his *Four Fundamental Concepts of Psychoanalysis*, Lacan explains this fear and desire in visual terms:

> From the moment that this gaze appears [the illusion of the self "seeing itself seeing itself," or the illusion of being gazed at by the Other], the subject tries to adapt himself to it, he becomes that punctiform object, that point of vanishing being with which the subject confuses his own failure. Furthermore, of all the objects which the subject may recognize his dependence in the register of desire, the gaze is specified as inapprehensible. (83)

The subject simultaneously desires (or desires to be) the object of the Other's gaze, and cannot apprehend the Other. Desire is always the desire of the Other, to merge with the Other (not the desire for the Other). The threat of the Other is not, then, to our identity, but rather to our perception of desire. A recognition of the Other creates a recog-nition of direct desire for an object. In his interpretation of this fear, Slavoj Žižek states:

> It is *they* who steal our enjoyment, who, by means of their excessive attitude, introduce imbalance and antagonism. With the figure of the Master, the antagonism *inherent* in the social structure is

transformed into a relationship of *power*, a struggle for *domination* between *us* and *them*, those who cause the antagonistic imbalance. (*Tarrying with the Negative* 210)[26]

Most followers of Lacan in literary-cultural studies characterize, as Lacan does, the relationship between the self and the Other as antagonistic, dichotomous, and active (even mimicry, as defined by Homi Bhabha, is antagonistic). We have already rehearsed the arguments against simple dichotomies. What happens, according to psychoanalysis, when the Other is expelled from the system? If the subject chooses not to explore this antagonistic relationship, then the subject is either psychotic or neurotic. In his explanation of the mechanism of psychoses, Lacan examines the way in which, in the absence of communication with the Other, the paranoiac views everything as speaking to him. Every sign has a profusion of signification. The relationship with the system, the outside, remains antagonistic, but now the entire world is conspiring against the paranoiac.

In *Titus,* Tamora's expulsion of Titus and Titus' expulsion of Aaron do not function in this manner. First, the expulsions are only from certain codes. Titus is still legible and present as is Aaron. Titus takes Aaron's word for the fact that his hand is a necessary sacrifice. While Tamora and Titus both internalize elements of what they have expelled, the paranoid does not. The paranoid, cut off from communication with the Other, in fact externalizes all signs; things speak to him and, in the famous case of Judge Schreber, he is connected to things. Schreber states repeatedly that "rays" from the outside bombard him and influence his actions. Even the actions of his therapist take on profound meaning for Schreber, but Schreber maintains a sense of self, albeit a confused one. The paranoid does not adopt the characteristics of the foreclosed Other. These characteristics are projected outwards. If the Lacanian Other were at work in the race relations in *Titus,* one would expect such projection.

Recognition of the workings of transversal power provides alternatives to psychoanalytic readings. Habib correctly points out "the desire and the danger" of the "master city" when faced by a native (91). Transversal theory posits that individuals move in subjective territories, which are, in turn, surrounded by official territories. The principle of pressurized belongings relates to a traditional notion of racism as a desire to keep "the other" from invading. As commonly used, the word "racism," however, implies only the resistance of the majority to the incursion of a minority marked by certain physical features. This does not account for the entirety of fears circulating among Shakespeare's audience. For instance, the battle of Ksar Kbir had political and religious overtones, yet its effect on England reached beyond these. Our use of the word "beyond" here can be understood with regard to subjective territory. Social, cultural, and political forces governing a specific subjective territory, typically as extensions of influential sociopolitical conductors, can acknowledge certain items as outside of an ideological framework and successfully protect the territory from threatening incursions.

The place of Aaron: too much pressure and the power of the fugitive

While Saturninus marries Tamora and refers to her "hue" as different from the Romans, the only character that is explicitly labeled as black must remain in the margins of the empire. It should be noted that, while *Titus*'s Rome marginalizes Aaron, the structure of the play itself does not. Aaron has more stage time than Saturninus and he is the only character who addresses the audience directly. Most of the major plot twists depend on Aaron's machinations. The well-documented success of the original productions of *Titus* supports the arguments already presented which demonstrate that the relationship between British citizens and the Blackamoor are more complex than mere exclusion or contamination. This conflict between interest in and fear of other groups can be seen in the artwork surrounding the play and the commentary on that artwork. Henry Peacham's 1595 drawing (Figure 8.1) represents the first scene of *Titus*.[27] The painting divides the players into three groups. The Romans stand on the left side of the painting, facing the Goths, who kneel close to the center and face almost directly back toward the Romans. Aaron stands by himself, beyond the Goths, facing almost full front. Richard Levin points out that Aaron should not have a sword in the picture: he should be bound, just as the Goths are (323). Aaron is positioned at the margins of the sketch, but, given that he is the only dark figure, and the only isolated figure, and the fact that he has a drawn sword, this position at the margins may serve to give Aaron more power. In the words of Geraldo U. de Sousa: "Peacham's Aaron is openly defiant while Shakespeare's character resorts to subterfuge and secrecy. Aaron of the play understands that he occupies the space at the margins of the Roman world, where there is no accountability, where acts of villainy are perpetrated with impunity" (104). Peele and Shakespeare's subtle Aaron is

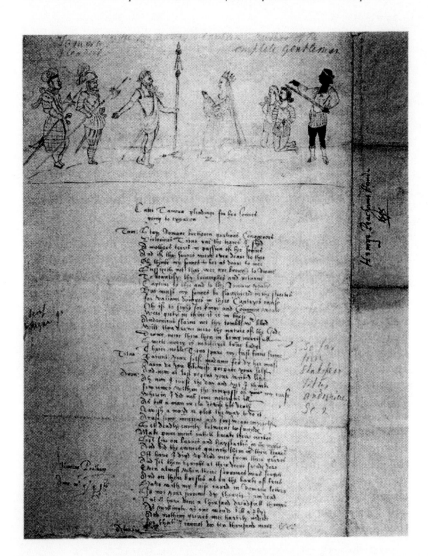

Figure 8.1   "Titus Andronicus" (drawing, Henry Peacham, 1595)

able to act with impunity precisely because he is at the margins. Peacham's sword-wielding Aaron is still at the margins and still free from bondage, but he also points, perhaps at Chiron and Demetrius. His machinations in the painting are not subtle. By pointing at Chiron and Demetrius, Peacham's Aaron clearly demonstrates the power he has in Rome. The contrast is striking between this Aaron and the Aaron in the ~1658 woodcut of the ballad *The Lamentable and Tragicall History*

Figure 8.2 "The lamentable and tragicall history of Titus Andronicus" (woodcut, 1658)

of Titus Andronicus (Figure 8.2). In the woodcut, Aaron is coal black, not wearing clothes, has exaggerated "wooly" curls, and is emerging from the ground of the painting. The stereotype that some scholars argue informs Peele and Shakespeare's writing of Aaron is clear, then, in this later woodcut, although not in the earlier drawing.

For the Elizabethan audience, in addition to being coded by his blackness, Aaron's ranting marks him as a becoming-standardized revenge tragedy villain, his invocation of devils marks him as a Moor, and his treatment of his child marks him as human. The "principle of translucency" is not just a statement of a multilayered character, but refers to a layering of individually legible codes represented on stage, in any social performance, or in any expressive medium; and through their simultaneous and related articulation, the codes create additional meanings and blendings of meanings, although not precise unity.[28] With the possible exception of Aaron's care for his child, nothing in his character is surprising. He has no depth beyond that which he needs to fashion devious schemes. Enacting multiple codes is not complexity of character; it is complexity of portrayal. The principle of translucency affects how the audience will interpret Aaron's actions, not how Aaron himself acts. In addition to the meanings of the respective codes, the principle of translucency suggests that the combination and blending of codes also has an effect. Why else would Peele and Shakespeare make the canniest character in the play perform such a multifaceted identity?

Habib explains Aaron as an example of reflexive "subtle, mimicking over-performance of his textual life to the point of making it meaningless and demolishing its ethical import" (98). In this reading, all of Aaron's lines about blackness are intended as an attack on the system that places the black body into the semiotic field of allegorical blackness. Aaron's jesting about deception while cutting off Titus's hand provides a clearer example of the character's use of sophisticated humor, and thus this explanation can hold. Aaron himself makes the first negative reference to racial markers: "My fleece of wooly hair that now uncurls, / Even as an adder when she doth unroll / To do some fatal execution?" (2.3.34–6). However, the second part of Aaron's line identifies the snake as both evil and female. Tamora, not Aaron, is about to enact her revenge, lying curled in wait to enact an execution. Aaron is not simply condemning Tamora, either for her "blackness" or her action. Aaron's opening monologue has linked Tamora to Semiramis, a racially marked Roman Empress who created a celebrated racially marked Roman dynasty. Herodian, at least, gives this dynasty credit for founding Western civilization. Habib asserts that such an allusion would not have been lost on the Shakespearean audience (93). The image, whether or not it serves as defensive mimicry of the white man's construction of blackness, also allows Aaron's identification with a Roman version of racial power, as embodied in Tamora. This is a primary example of the conscious assimilation of codes. Aaron affiliates Tamora with Roman history and himself with the feminine in Tamora in order that he himself might gain the Roman affiliation.

Habib's argument about Aaron's adoption of the role being a mockery of the white man's stereotypes can explain many of Aaron's rants, including his extended monologues in the final act of the play. In this final act, Lucius refers to Aaron as an "inhuman dog, unhallowed slave" (5.3.14), again mixing potential racial slurs with religious overtones. When Aaron is sentenced to die, Aemilius refers to his crimes rather than his skin tone. It is Aaron himself who refers to blackness having caused his evil. Such bitter mockery of the Roman prejudice seems oddly out of place in a man seeking the aid of Romans. In fact, Aaron refers to his color only once in the extended monologues: "like a black dog, as the saying is" (5.1.122). Habib connects this saying with Aaron's statement, in reference to his child, that, "Coal black is better than another hue / in that it scorns to wear another hue" (4.2.99–100), claiming that both statements represent black pride (107). The two statements differ slightly in their relationship to the idea of crossing over and dissembling. Furthermore, for mimicry to be effective, Aaron would need to perform such a mockery of codes before the audience of the Andronici. In this reading, Aaron enacts an elaborate deceit conceit, convincing the Andronici of his nature as a barbarously evil black moor only in order to later catch them in a trap. In other words, according to Habib, Aaron chooses to adopt the codes with which he feels he may do the most damage to the Andronici; Aaron's deceit conceit is hiding behind the images that others have constructed of him. He does snare the Andronici, but this is because he plays the part of the noble Roman when in front of the Romans. When Aaron reinforces his blackness, he does so in private, or in the presence of Tamora, yet not in front of Saturninus or the Andronici.

If mimicry is not the answer, why then would Aaron reinforce his identity as black and evil? The rants are characteristic of the hybrid vice/villain figure, but Peele and Shakespeare created a canny figure rather than a parodic villain. In this final monologue, Aaron separates himself from the Roman ideal by citing an inability to display shame. Titus twice asserts that the Roman ethic demands the death of a child to prevent the shaming of a Roman name. Saturninus would certainly view Aaron's child as an insult, but, potentially, the affirmation that Aaron's son is indeed wholly black, and therefore unable to cross into Roman society, allows Saturninus to spare him. The child is not a threat. Chiron and Demetrius have already worried that the child might usurp them, although Aaron makes it clear that he is black. Demetrius and Chiron are pretenders at being white – "white-limed" (4.3.98) – and have access to Roman power: the child does not have this access.

Despite Aaron's negative associations with darkness, he also refers to the "pale envy" of others (2.1.4), and Demetrius remarks with fear that his mother looks "pale and wan" (2.3.90), which both suggest that Rome's lack of pigmentation does not universally associate them with goodness. Given that the later of these two lines immediately follows the images of Saturninus as snow-white horse, Aaron as raven-dark lover, and Tamora's virtue becoming spotted, this further reference to whiteness within the same scene must have some racial connotation. If race is, as we think, a semiotic category, then, perhaps, in the semiotic field with "whiteness" is a demonstrated feebleness, the opposite of the lustful Goths. In addition to the potential irony of Demetrius' lines (based on the fact that it is Lavinia and Bassianus who are in danger rather than Tamora), he could be expressing some concern with his mother's assimilation into the Roman system. In either case, Tamora's response expresses clearly her contempt for Roman virtue and honor. She falsely accuses Lavinia and Bassianus of leading her to the mouth of hell to kill her. Her alleged fear and paleness come from this untruth. Tamora has nothing to fear, and thus does not, in all like-lihood, look pale, which adds credence to the above reading of the lines. Marcus speaks of "The proudest panther" (2.2.21), which is viewed as worthy sport for the king. Later Peele and Shakespeare explicitly link Aaron with the black panther. Of course, this still equates Aaron with a beast; here he is not a fly, crushed with impunity; rather the image, when Marcus invokes it, is one of a noble beast set to ascend "the highest promontory top" (2.2.22). Again, the color line and the field of terms associated with color depend on their context relative to Roman tradition and the power hierarchy. Aaron makes use of others' perceptions of his blackness.

While the image of Aaron as a panther leaves some room for ambi-guity in its relationship toward racial attitude, Bassianus's statement does not: "your ["swart"] Cimmerian / Doth make your honor of his body's hue, / Spotted, detested, and abominable" (2.3.72–4). Clearly Bassianus does not approve of Aaron's skin tone, especially when contrasted with Saturninus' "snow-white goodly" nature (2.3.76). Bassianus refers to Aaron's spottiness, not his blackness. Again, the fear of crossing over makes Aaron, as Moor, too close to the Roman white-ness. If, indeed, Aaron is spotted with marks of both types of colora-tion, then he is near enough to the boundary to corrupt the Queen of Rome. Bassianus states, though, that it is Tamora's honor that is in danger and not her body. By extension, Aaron sullies the honor of Rome. Nonetheless, it is Bassianus whose honor is most sullied in this scene as he dies out of favor with the Roman empire, if only in effect of Aaron's machinations. Again, in expelling Aaron's spottiness, Bassianus himself ends life in a "stained hole" (2.3.210).

Unlike Chiron and Demetrius, Aaron and Tamora are both bestowed a single moment of humanness. Both care deeply about their children. When Tamora's maternal instincts, Aaron's protective nature, and Titus' slaughter of his two children are juxtaposed, another aspect of the Roman virtue of the play comes into light. Aaron receives the due punishment for not being a good Roman, but manages to outlive Titus based, in part, on the utility of the child whose life he saves. The play does not explicitly deal with the fate of the child, repeatedly marked as black by those who see him. The child's blackness, in fact, displays the Elizabethan attitude toward the dominance of blackness in racial mixing, or the ability of the black man to corrupt whiteness as outlined by Loomba. In other words, this child embodies one of the greatest racial fears of the time, and, yet, technically, the child is the offspring, albeit illegitimate, of the Roman empress, although she may not play her role well, and so the child lives. Aaron's line continues on, just as Titus' does. The child embodies a mixing of codes: he will presumably be raised in accordance with Roman religion, but is from Moorsih heritage. He is Roman, by virtue of having been born to the empress. The child represents villainy, but is not himself a villain. As noted above, each code carries with it corresponding power structures. The child pulls Aaron out of the villain's role and makes him subject to, if not a subject of, Roman law. The child assimilates all codes and is simultaneously expelled by them. In private, Aaron refers to the child as "tawny," explicitly defined as an in-between state for Elizabethans (Forbes 123). To be tawny is to defy other classifications, to be neither wholly one nor another; it is to be, effectively, in a constant process of becomings and comings-to-be.

The position at the margins of identity, as already discussed, is not always at the margins of power. The Moor, and therefore Aaron, given the way in which he enacts multiple codes, and given the fluidity of the Moorish code he enacts, might actually be the site of the greatest resistance in *Titus*. He can simultaneously or sequentially adopt or shrug off codes. He would not be able to erase the markers of difference that others see, but he does manage to integrate himself into the Roman court. Reynolds explains this kind of power in relationship to gypsies in early modern England: "The signifying modes combined with the concept of gypsyism to produce a different sign, one that was more ideological and abstract and consequently obscured or displaced the expected referent. ... Consequently, the gypsy sign became a sociopolitcal conductor whose transversal power and objective agency made it an unusually trenchant mechanism for the expansion of transversal territory" (Reynolds, *Becoming Criminal* 25). Objective agency refers to "the power of a sign to frustrate or transcend the effectual parameters and expectations generally ascribed to that sign by a system of codification within which it functions" (24). Fugitive exploration creates spaces in which objective agency can be obtained. Aaron, as a fugitive figure enacting multiple codes, provides a perfect example of the simultaneous power and limitations of fugitivity.

Some concluding, opening remarks

If *Titus* is indeed a play about alterity, as Loomba asserts, "race has functioned as one of the most powerful and yet most fragile markers of social difference" (203). "Race" in *Titus* involves more than mere skin tone, and while the negative imagery associated with darkness permeates the play, attitudes toward filial duty to both family and country supplement the concept of race. To say that Peele and Shakespeare present Aaron's evil as motivated by "race" is to ignore historical context, to ignore the intricacies of the text, and to put a modern focus on a relatively small group of lines within the body of the work. A simple count of lines in the play will show the relative importance of the Roman ideal to the racial markings. Peele and Shakespeare tie the idea of racial identity contextually not only to the concepts of good and evil, but also directly to honor, justice, and the other qualities that make up the Roman ideal. Peele and Shakespeare mark Aaron and Tamora as other, as outside, and Titus is likewise marked and forced to the margins. All three remain outside of the proper Roman ideal. It is the process by which people can become other than they are or appear to be, or be forced to come-to-be different, and the degree to which this occurs, which is at stake in the play. This question of difference and boundaries is one of the surprisingly modern features of *Titus*. Etienne Balibar speaks of a neo-racism which "is not biological heredity but the insurmountability of cultural differences, a racism, which, at first sight, does not postulate the superiority of certain groups or peoples in relation to others but 'only' the harmfulness of abolishing frontiers" (21). While perceived biological difference is at stake in *Titus*, the idea had already become entangled with issues of cultural and national difference in the articulatory spaces and formations (in art, literature, theology, and politics) dealing with social identity during Elizabeth's reign, which, as we have seen, are engaged by and integral to the plot of *Titus*.

The play does not provide an easy pattern as it joins every instance of expulsion with one of assimilation. Ignoring current commentary on the quality of the text, *Titus* was not a failure. It was one of the most popular plays of its period. If the task of the play was to provide a means of identifying the cultural codes of a new Britain, then the principle of pressurized belongings explains the play's inability to meet its task. If expelling someone from a code requires the contamination of the expeller, then Titus, as protector of Rome, was doomed from the start, as was the project of the British empire. The play demonstrates the related processes of becomings and comings-to-be, and the potential for people, as represented by the characters, to negotiate subjective and official territories, and thus evinces the dynamism of pressurized belongings. Every character crosses over a boundary: *Titus* is not a roadmap to empire; rather, it suggests modes of interaction that will manifest as the empire evolves.

Notes

1. For a discussion of Peele's collaboration with Shakespeare on *Titus* and the date of production, see Gary Taylor, *Buying Whiteness: Race Culture, and Identity from Columbus to Hip-hop* (Palgrave Macmillan, 2005), 204.
2. On the "principle of translucency," see Donald Hedrick and Reynolds, " 'A little touch of Harry in the night': Translucency and Projective Transversality in the Sexual and National Politics of *Henry V*," in Reynolds, *Performing Transversally:* 171–88.
3. On the theory, methodology, and aesthetics of transversal poetics – beyond this book, see the following works by Bryan Reynolds: "The Devil's House, 'or worse': Transversal Power and Antitheatrical Discourse in Early Modern England" (*Theatre Journal* 49.2 [1997]: 143–67); *Becoming Criminal: Transversal Performance and Cultural Dissidence in Early Modern England* (Baltimore: Johns Hopkins University Press, 2002), 1–22; *Performing Transversally: Reimagining Shakespeare and the Critical Future* (New York: Palgrave Macmillan, 2003), 1–28.
4. See F. P. Wilson, *The English Drama: 1485–1585*, for a complete index of play titles and publication dates. According to Elliot Tokson, even *Othello*, Shakespeare's supposed overthrow of stereotypes, follows only six non-Shakespearean portrayals of black men on stage.
5. We use the phrase "revenge tragedy" to denote only those tragedies "whose leading motive is revenge and whose main action deals with the progress of this revenge, leading to the deaths of the murderers and often the death of the avenger himself" (Bowers 62). This school is commonly considered to have originated with Kyd (see Bowers, Bevington). Earlier Elizabethan adaptations of Senecan drama, such as *Gorboduc* (1562), take politics and decorum as their main focus, rather than revenge (Bowers 64). According to Bowers, this form may have been transmitted through the Italian neo-Senecan tradition, although these plays also did not focus on revenge. Arthur Freeman associates the development of the Machiavellian villain with the development of revenge tragedy; the villain of the revenge tragedy must create elaborate plots in order to fool the protagonists. It is the convolutions of these plots that lead to the eventual deaths that make the play tragic (1–24).
6. For specific arguments on the dating of the plays, see Freeman, *Thomas Kyd: Facts and Problems* (1–24).
7. The character, Revenge, also appears onstage in Kyd, which is one reason that scholars consider *The Spanish Tragedy* to be the transitional play between religious drama and revenge tragedy. The character "Revenge" also rants, just as his predecessor vice figures did. During his rants, however, revenge cosigns himself to the role of chorus. He watches and comments on the onstage events, but does not participate actively in them. In addition, Revenge is assisting Andrea, who is arguably the protagonist figure of the text, although in the end Andrea and Revenge decide to reunite both "friends and foes" in Hades. This ambiguous attitude toward revenge, the idea that revenge could go too far, remains consistent through the entire genre. The villains are Machiavellian and the heroes are generally flawed in some manner.

8. In his final scene, Barabas, the Jew of Malta, remarks: "Know, governor, 'twas I that slew thy son / I fram'd the challenge that did make them meet: / Know, Calymath, I aim'd thy overthrow: / And, had I but escap'd this stratagem, / I would have brought confusion on you all, / Damn'd Christian dogs, and Turkish infidels!" (5.5.80–4). And Villuppo, from *The Spanish Tragedy*, states: "Thus have I with an envious, forged tale/Deceived the king, betray'd mine enemy./And hope for guerdon of my villainy" (1.3). Unlike Barabas, Villuppo gives no reason for his evil, but delights in perpetrating it.

9. Maley cites the chorus in *Henry V* as an example of the conflation of England and Rome: "There are at least three different kinds of history at work here. Henry's return to London is first compared with Caesar's to Rome, then, less enthusiastically, with Essex's anticipated arrival from Ireland" (35). Given that *Titus* is set in Rome, it connects Rome to contemporary Britain in a much simpler manner, without the intermediary of another era of British history.

10. For a detailed argument about the connection between Machiavelli and Shakespeare, see John Roe's *Shakespeare and Machiavelli* (New York: D. S. Brewer, 2002).

11. The connection between Moor and blackness in English language develops in the late 1500s. Minsheu distinguishes between various types of Moors, Moors that have become Christian, Ethiopian Moors, light-skinned Moors, Arabs, and dark skinned Africans (Forbes 84). It was possible to have dark skin and be a Christian but still to be referred to as Moorish. A light-skinned Arab could be called Moorish, as could a native American or Indian. Essentially, as the political situation between Africa, Britain, and America altered, the definitions of Moor and negro altered along with it. The connection and distinction between "black," "negroe," "slave," and "Moor" does not develop fully until the 1700s during the height of the North Atlantic slave trade. At this point, Moor came to refer to North African, while black and negroe became almost synonymous. In the aftermath of Ksar Kbir, however, the word "Moor" also implied non-Christian. We employ the word "Moor" here to mean a practicer of Islam. The word itself, as has been well documented by Julia Lupton, Bernard Harris, Jack D'Amico and others, comes to stand in for a variety of race, class, national, and religious differences. Again, however, it is necessary to note that Shakespeare enters the theater as such linkages were being established, as seen by the fluidity with which characters switch their labels of Aaron.

12. According to the *Riverside Shakespeare*, Shakespeare may have written *Titus* as early as 1592. Henslow recorded a production of *Titus & Ondronicus* on January 23, 1594. In either case, *The Battle of Alcazar* had already presented a Moorish villain in the dramatization of the battle of Ksar Kbir and gained the approval of audiences.

13. Barthelemy states that Muly Mahamet, the villain of *The Battle of Alcazar*, was the first of the emerging villain types to be Moorish. It is not clear from textual evidence whether Muly Mahamet was actually black as he is the son of a white moor and a black woman (see endnote 3).

14. As noted by Barthelemy, in *Tamburlaine*, Parts I and II, various Moorish royals pay tribute to Tamburlaine. Other Moors in contemporary pieces "seem to be there as exotic personages to give substance to the exotic

names the conquering kings like to bandy about" (76). Characterizing Moors as exotic is not the same as characterizing them as villains.

15. Actually, while Harris states this date as fact, D'Amico provides evidence to support 1548 as the beginning of formal trade. The documents most frequently cited in these debates are various editions of Hakluyt's *Principal Navigations, Voyages, Traffiques & Discoveries of the English Nation* and the work of Leo Africanus. Emily Bartels provides an excellent summary of Hakluyt and Africanus in her essay "Making More of the Moor." As all positions within the debate place this formalized trade before *Titus*, the nuances of the historical argument are not important for the purposes of our analysis. Instead, we intend to highlight the confluence of the battle of Ksar Kbir, the sudden appearance of actual "black" men in England, the formation of the revenge tragedy, and the formalization of relations with Barbary.

16. We return to the difficulty that began our argument. What "ism", precisely, is governing the Elizabethan response to Moors? It is not precisely racism, as Moors remain distinct from "blacks" during some writings of the time. It is not just anti-Islamic sentiment, as seen by the religious terms of the alliance. The label of barbarian was perhaps the most consistently applied, but Elizabethans applied this label liberally to many groups of foreigners. Lupton suggests "a 'culturalist' rather than biologistic ordering of intergroup relations" (73). Lupton demonstrates a relation between Pauline doctrine of "division of nations" and fear of the Moor (not of the racially marked man), which she cites as central to *Othello* (74). We agree with the importance of religious difference in *Titus*, but to this religious difference we add perceived biological difference (which is emergent, not formed), cultural difference in terms of nationality, and the cultural marker of villain.

17. For Barbara Everett, the literary antecedent of *Othello* comes from Spain, which places the Moor in yet another context. Everett attempts to dismantle the "too simple 'African' sense of *Othello*" by showing the connections between the Spanish Moors and England. Her arguments are helpful to formulating an understanding of *Titus* only inasmuch as they further complicate the notions of a simple relationship between black and Moor.

18. Engel and Reynolds, respectively, more accurately associate the fluidity of identity with figures like Moll Cutpurse in Middleton and Dekker's play *The Roaring Girl*, who exist in the liminal spaces of sexual-gender identity. See Reynolds, *Becoming Criminal*, 24–64.

19. Other examples of this type of work include Wole Soyinka's enigmatic, tongue-in-cheek look at the politics of non-traditional histories of Shakespeare in relation to Arab elements within the text of the plays, see Wole Soyinka, "Shakespeare and the Living Dramatist," in Catherine M. S. Alexander and Stanley Wells, eds. *Shakespeare amd Race* (Cambridge, UK: Cambridge University Press, 2000), 82–100. Lupton, "Othello Circumcised," which examines the degree to which physical markers of Moorishness are important textually and historically; and Eldred Jones, in his catalogue of literary and historical data about Shakespeare's treatment of Moorishness in which he locates "the triumphant genius of Shakespeare" (132).

20. We choose to avoid excessive verbiage and semantic acrobatics here. No one word or simple phrase can stand in for "These" figures. "Othello's Countrymen" implies that nationalism is the primary concern. "Racially marked"

assumes that the physical markers of race are primary. The word "marginalized" does not take into account the central position that these figures take in many later plays, nor the centrality of trade relationships that England developed. We have already explained the difficulty with the words "black" and "Moor" and their relationships to the word "villain." For reasons that will become apparent as our argument progresses, the term "Other" cannot be judiciously applied. "These figures" are then the perceived-as-black-and biologically different (except when politically convenient), potential convert/heathens (except when compared to Catholic-outsiders; unless they had trade goods), interesting to watch on stage, not necessarily, although likely, villains-barbarians who had civilization, potential British citizens, but still non-English threats to national identity. Hereafter, we can simply refer to "Aaron."

21. For more on "articulatory spaces" and "Shakespace," see Reynolds, *Performing Transversally*, 1–28 and Hedrick and Reynolds, "Shakespace and Transversal Power," in *Shakespeare Without Class: Misappropriations of Cultural Capital*, Eds. Bryan Reynolds and Donald Hedrick (New York: Palgrave Macmillan, 2000): 3–47.

22. For more on Hegel's concept of becoming, see his *Phenomenology of Spirit*, trans. A. V. Miller (Oxford: Oxford University Press, 1997): 10–14.

23. See Gilles Deleuze and Félix Guattari, *A Thousand Plateaus: Capitalism and Schizophrenia*, trans. Brian Massumi (Minneapolis: University of Minnesota Press, 1987), especially the chapter "Becoming-Intense, Becoming-Animal, Becoming-Imperceptible ..." (232–309).

24. See Reynolds, *Performing Transversally* (1–29).

25. Jean Laplanche says of the Lacanian Symbolic, "to attempt to contain the meaning of the 'Symbolic' within strict boundaries – to define it – would amount to a contradiction of Lacan's thought, since he refuses to acknowledge that the signifier can be permanently bound to the signified" Jean Laplanche and J.-B. Pontalis, *The Language of Psychoanalysis*, trans. Donald Nicholson-Smith (New York: Norton, 1973), translation of *Vocabulaire de la Psychanalyse* (Paris: Presses Universitaires de France, 1967), 440.

26. For a detailed analysis of the use of psychoanalytic theory in race studies, see Simon Clarke *Social Theory, Psychoanalysis and Racism* (New York: Palgrave Macmillan, 2003).

27. In 1999, June Schluter published "Rereading the Peacham Drawing" in *Shakespeare Quarterly 50*. This piece argues that Peacham's drawing is, in fact, not of Shakespeare's *Titus Andronicus*, but rather of the German *Eine Sehr Klagliche Tragedia Von Tito Anronico und der Hoffertigen Kayserin*. Richard Levin argues against this interpretation of the drawing. For the purposes of our examination of the multiple codes involved in the presentation of ethnicity, religion, etc. it is not necessary to determine to which *Titus* the drawing refers. Schluter closes her argument by stating that "in the Spring 1999 issue of *Shakespeare Bulletin*, Herbert Berry offers what will surely become the definitive reading of the elusive date ... 1594" (184). Thus, regardless of which play Peacham drew, scholars are in agreement that the drawing is English and roughly contemporary with *Titus*.

28. On the "principle of translucency," see Donald Hedrick and Bryan Reynolds, " 'A little touch of Harry in the night': Translucency and Projective Transversality in the Sexual and National Politics of *Henry V*," in Bryan Reynolds, *Performing Transversally*: 171–88.

Works cited

Alexander, Catherine and Stanley Wells Eds. *Shakespeare and Race*. Cambridge: Cambridge University Press, 2000.

Babula, William. *Shakespeare in Production, 1935–1978: a Selective Catalogue*. New York: Garland Publishing, Inc., 1981.

Balibar, Etienne. "Is there Neo-Racism?" *Race, Nation, and Class*. Trans. Chris Turner. London, 1991.

Barthelemy, Anthony Gerard. *Black Face Maligned Race*. Baton Rouge: Louisiana State University Press, 1987.

Bartels, Emily C. "Making More of the Moor: Aaron, Othello, and Renaissance Refashioning of Race." *Shakespeare Quarterly* 41.4 (1990).

Bowers, Fredson. *Elizabethan Revenge Tragedy: 1587–1642*. Gloucester, Mass: Peter Smith, 1959.

D'Amico, Jack. *The Moor in English Renaissance Drama*. Tampa: University of South Florida Press, 1991.

De Sousa, Geraldo U. *Shakespeare's Cross-Cultural Encounters*. New York: Macmillan – now Palgrave Macmillan, 1999.

Engel, William. "Death Slips onto the Renaissance Stage," in *Medieval and Early Renaissance Theatre and Drama*, ed. Laurie Postlewate (Amsterdam: Rodopi, forthcoming 2006).

Everett, Barbara. " 'Spanish' Othello: the making of Shakespeare's Moor." *Shakespeare and Race*. Ed. Catherine Alexander and Stanley Wells. Cambridge: Cambridge University Press, 2000. 64–81.

Forbes, Jack D. *Africans and Native Americans: the Language of Race and the Evolution of Red-Black Peoples*. Chicago: University of Illinois Press, 1993.

Freeman, Arthur. *Thomas Kyd: Facts and Problems*. Oxford: Clarendon Press, 1967.

Fredrickson, George. *The Arrogance of Race: Historical Perspectives on Slavery, Racism, and Social Inequality*. Middletown, CT: Wesleyan University Press, 1988.

Habib, Imtiaz. *Shakespeare and Race*. New York: University Press of America, Inc., 2000.

Hadfield, Andrew. "Bruited abroad: John White and Thomas Harriot's Colonial Representations of Ancient Britain." *British Identities and English Renaissance Literature*. Ed. David Baker and Willy Maley. Cambridge University Press: Cambridge. 159–77.

Hall, Kim. *Things of Darkness: Economies of Race and Gender in Early Modern England*. Ithaca: Cornell University Press, 1995.

Harris, Bernard. "A portrait of a Moor." *Shakespeare and Race*. Ed. Catherine Alexander and Stanley Wells. Cambridge: Cambridge University Press, 2000. 23–36.

Hegel, G. W. F. *Phenomenology of Spirit*. Trans. A. V. Miller. Oxford: Oxford University Press, 1997.

Hunter, G. K. "Elizabethans and Foreigners," *Shakespeare and Race*. Ed. Catherine Alexander and Stanley Wells. Cambridge: Cambridge University Press, 2000. 37–63.

Jones, Eldred. *Othello's Countrymen: the African in English Renaissance Drama*. London: Oxford University Press, 1965.

Knapp, Jeffrey. *Shakespeare's Tribe*. Chicago: University of Chicago Press, 2002.

Lacan, Jacques. *The Seminar XI, The Four Fundamental Concepts of Psychoanalysis*. Ed. Jacques-Alain Miller, trans. Alan Sheridan. New York: W. W. Norton & Co., 1977.

—— *The Seminar, Book III. The Psychoses*. Ed. Jacques-Alain Miller, trans. Russell Grigg. New York: W. W. Norton & Co., 1993.

—— *Ecrits: a Selection*. Trans. Alan Sheridan. New York: W. W. Norton, 1997.

Levin, Richard. "The Longleat Manuscript and *Titus Andronicus*." *Shakespeare Quarterly* 53:3 (Fall 2002), 323–40.

Little, Arthur. *Shakespeare Jungle Fever: National-Imperial Re-Visions of Race, Rape, and Sacrifice*. Stanford: Stanford University Press, 2000.

Loomba, Ania. " 'Delicious Traffick': Racial and Religious Difference on Early Modern Stages." *Shakespeare and Race*. Ed. Catherine Alexander and Stanley Wells. Cambridge: Cambridge University Press, 2000. 203–24.

Lupton, Julia. "Othello Circumsized: Shakespeare and the Pauline Discourse of Nations." *Representations* 57, 1997.

Maley, Willy. "This Sceptred Isle," *Shakespeare and National Culture*. Ed. John J. Joughin. New York: Manchester University Press, 1997. 83–108.

Reynolds, Bryan. *Becoming Criminal: Transversal Performance and Cultural Dissidence in Early Modern England*. Baltimore: Johns Hopkins University Press, 2002.

—— *Performing Transversally: Reimagining Shakespeare and the Critical Future*. New York: Palgrave Macmillan, 2003.

—— and Joseph Fitzpatrick (with Janna Segal), "Venetian Ideology or Transversal Power? Iago's Motives in the Means by which Othello Falls," in *Performing Transversally: Reimagining Shakespeare and the Critical Future*. New York: Palgrave Macmillan, 2003. 53–84.

Roe, John. *Shakespeare and Machiavelli*. Cambridge: D. S. Brewer, 2002.

Shapiro, James. *Shakespeare and the Jews*. New York: Columbia University Press, 1996.

Spivack, Bernard. *Shakespeare and the Allegory of Evil*. London: Oxford University Press, 1958.

Thompson, Ayanna. "The Racial Body and Revenge: Titus Andronicus," *Textus* 13: 325–46. 2000.

Tokson, Elliot H. *The Popular Image of the Black Man in English Drama: 1550–1688*. Boston: G. K. Hall and Co., 1982.

Žižek, Slavoj. *Tarrying with the Negative: Kant, Hegel, and the Critique of Ideology*. Durham, NC: Duke University Press, 1993.

9
Awakening the Werewolf Within: Self-help, Vanishing Mediation, and Transversality in *The Duchess of Malfi*

Courtney Lehmann & Bryan Reynolds

When thou killed'st thy sister,
Thou took'st from Justice her most equal balance,
And left her nought but her sword.
(John Webster, *The Duchess of Malfi* 5.5.39–41)

The mediator between power and freedom,
the redeemer without which power remains
violence and freedom caprice, is therefore – love.
(Richard Wagner qtd. in Deryck Cooke,
I Saw the World End, 17)

Self-help's popularization by late capitalism acknowledges the extent to which our relationships can be mediated through commodities, yet overstates its potential to mediate satisfactorily our sense of our "inner" selves. But to what extent were the mechanisms of self-help already in place within the emergent capitalism of early modern culture?[1] According to the entertaining speculations of the 1998 Academy Award winning film, *Shakespeare in Love*, even Will Shakespeare sought the help of an apothecary-therapist to help him access the "poet of true love" that lay dormant within his emasculated body.[2] Nevertheless, the character most in need of therapy in this film is not Shakespeare but one of his fellow playwrights, John Webster, the vagrant boy who hangs around the playhouse, tortures mice, witnesses forbidden sexual trysts, and revels in bloody endings, as he explains in response to the Queen's inquiry as to whether or not he enjoyed Will's production of *Romeo and Juliet*: "I liked it when she stabbed herself, your majesty." What we know of the real John Webster does little to mitigate this image of the disturbed pre-teen projected by *Shakespeare in Love*, for Webster's dedicatory prefaces to readers and patrons are rife with images of self-deprecatory language and negativity. In his preface to *The White Devil*, for instance, Webster offers the following caveat: "If it be objected this is no true dramatic poem, I shall easily confess it; *non potes in nugas dicere plura meas: ipse ego quam dixi*" (ln. 15–16), which translates as, "you cannot say more against these trifles of mine than I have said myself." Elsewhere, the brooding, melancholy tragedian envisions as grim a scenario as his own death, likening the pages of his book to "winding sheets," or shrouds (*Duchess* ln. 19), in a gesture of literal self-abnegation.

Webster may have benefited enormously from Anthony "Tony" Robbins's best-selling book, *Awaken the Giant Within: How to Take Immediate Control of Your Mental, Emotional, Physical, and Financial Destiny*, a self-help publication that illuminates an escape from the abyss of self-degradation. Whereas Shakespeare is cited throughout the book as a positive frame of reference for this process of awakening, Webster, alas, is not quoted once. A reason for this might be that Webster had already fashioned, like Robbins, his own cult of self-help in *The Duchess of Malfi*, one that shows how, in times of crisis, people place "giants" at the margins of what transversal theory calls our "subjective territory": the emotional-conceptual range from which individuals – subjected by a particular ideology – perceive and experience their environment.[3] As a result of their positioning at the periphery, the "giants" are not seen for what they often are: emanations of the "werewolf" within and, indeed, "vanishing mediators" in a process whereby self-help can revert to totalitarianism.

Here, we adapt Fredric Jameson's understanding of vanishing mediators as certain kinds of dialectical, intermediary structures, sociopolitical entities, such as a familial, religious, educational, juridical, or governmental institutions, that precondition their own transubstantiation, thereby allowing new societal structures to manifest themselves.[4] In transversal terms, such entities, which may be influential individuals as well as institutions, are called "sociopolitical conductors." These are the mental and physical movers, orchestrators, and transmitters that promote or oppose partially or predominantly, and often contradictorily, the dominant ideology of the society in which they operate. The sum of these constitute "state machinery," a concept that accounts for the singular and plural, human and technological influences that work tirelessly but typically futilely to manufacture societal coherence and symbiosis. Therefore, in broadening Jameson's concept of the vanishing mediator, as Slavoj Žižek does,[5] to apply to both specific individuals and groups, we recognize how vanishing mediators facilitate transition into expansion, reconfiguration, and novel organization of a subjective territory, the official territory of which they are a part, and/or the society that created them. Moreover, the special kind of sociopolitical conductor or signifier with the "objective agency" to conduct thought and behavior[6] that becomes a vanishing mediator (and sometimes also a *medium*) is often speculative and extraordinary, and thus works to negotiate social fantasy, typically a fantasy that contributed to its production in the first place.

The Duchess of Malfi presents a fantastic relationship between lycanthropy, also known as werewolf syndrome,[7] and its representation of the sociopolitical category "woman" as the vanishing mediator that is the conductor through which Ferdinand's self-reformation is able to occur.[8] Specifically, we are interested in exploring how Webster exposes the mechanisms whereby woman is elevated to the status of a traumatic objective agent, a "Thing" – the Freudian *das Ding* – only as the means to a far more damning revelation: that the excesses historically ascribed to woman expose the failures of patriarchy to confront its constitutive lie, that is, its failure to see itself as anything other than a hopelessly flawed compromise between the indulgence and prohibition of pleasure. With our fugitive exploration of *Duchess*, we hope to reveal Webster's depiction of Ferdinand's descent into lyncanthropy as a nihilistic form of self-help that transverses a fantasy of the woman-as-thing to arrive triumphantly at a subjective evacuation bountifully imbued with the interactive processes of "becomings" and "comings-to-be," whereby people incorporate conceptually, emotionally, and/or physically the traits of others either willfully and/or passively.[9]

Whereas much critical ink has been shed concerning Ferdinand's progressive lapse into perversion in *Duchess*, little has been said about his crisis of preferment as a potentially mitigating factor.[10] As the younger brother, Ferdinand would normally be disposed to pursue preferment via the Church, owing to the restrictions posed by the laws of primogeniture. However, in *Duchess*, the eldest son is also a Cardinal – and a lascivious, acquisitive one at that – who leaves Ferdinand with no hope of competing with big brother as the figure who has monopolized both the secular and sacred playing fields. Despite being hailed as the "Great Duke of Calabria," Ferdinand, as his affliction with melancholy suggests, lacks what Tony Robbins calls a consistent "Master System" of reference-points for his identity. Ferdinand's first opportunity to carve for himself a stable subjective territory within the symbolic order is the familial crisis posed by the widowing of his sister, which presents him with a proving ground for besting his brother and redeeming his compromised masculinity. Were Ferdinand to act in the best interest of the Church, which considers second marriage adulterous, he would respond to the specter of his sister's suddenly unmoored desire by urging her not to marry; however, as Duke, his most shrewd course of action would be to marry her to a man who would enhance the power-base of the family amidst the contested principalities of feudal Italy.[11]

Ferdinand's initial solution to this dilemma is a compromise-formation that exposes the degree to which his recommendation of abstinence is permeated by the specter of his own sexual desire.[12] Indeed, a variation on the Oedipal triangle can be seen in Webster's depiction of the Cardinal as a kind of father-figure, the Duchess as a mother-substitute, and Ferdinand as the melancholy son-in-crisis who exhibits inordinate interest in his sister's sexuality once he is presented with unmediated access to her in the wake of her husband's death. Interestingly, Ferdinand's attempt to negotiate this crisis reveals the perverse compensatory mechanisms of self-help, which, as we shall see in this case, almost invariably abandon the parameters of self-conception in the interests of sociopolitical damage control, which is to say, in the interests of the overarching official territory of which Ferdinand's subjective territory is merely a component. In fact, the unstated objective of self-help in both *Duchess* and Robbins is to create what Theodor Adorno refers to as a deindividualized "post-psychological" subject (152), and, at last, as Žižek would insist, of a totalitarian society (Žižek, *Metastases* 7–28).

Ferdinand is a figure who, according to Robbins's rhetoric, might benefit from awakening his inner giant through the power of "neuro-associative conditioning." The goal of neuro-associative conditioning is to negotiate the utilitarian calculus, that is, the ratio of pain to pleasure, in a way that "create[s] a direct highway out of pain and into pleasure with no disempowering detours" (124). The first step in this process is to establish a system of referents, a set of "props" for our identity that will help shore up the boundaries of our subjective territory and keep it from collapsing into the fearful void posed by an "identity crisis" – what Robbins refers to as "the ultimate pain" (420) – so that we can gain the agency necessary to become what we want. Read in the context of the historical controversies that form the backdrop of Jacobean tragedy, Robbins's system might be interpreted in light of "Catholic" doctrine, for Robbins assumes a certain degree of transubstantiation between subject and referent as a means of mediating

the anxiety posed by this identificatory gap. The more external referents or props we provide our identity, according to Robbins, the more
stable, consistent, and reliable it becomes, so that one may purpose-
fully become other; hence his assertion that "[t]he power of reading a
great book," for example, "is that you start thinking like the author.
For those magical moments while you are immersed in the forest of
Arden, you are William Shakespeare" (402). In transversal terms, by
moving "subjunctively," exploring the "as ifs" and "what ifs" that
make such reading special, one can preserve or augment agency and
embark on empowering becomings.

The point at which the discourse of Robbins's self-help and
Webster's play part ways is the moment when the charming forest of
Arden becomes a traumatic wilderness, a void that is *con*substantial to
Being, which cannot, therefore, be bridged, but, may at best, be tra-
versed. Should this occur, and a subject experiences alienation, adventure, or danger – what we call "wilderness effects" – perhaps the subject
will come-to-be something else, something potentially unrecognizable
within the perceptual schemes of official territory. For if the goal of
standard self-help methodology is to provide agency and reinforce subjectivity at all costs, then the objective of *Duchess* is quite the opposite:
to open up "the most radical dimension of subjectivity itself," what
Žižek would call – looking through the lens of his own obsession with
lack – "the Void which 'is' the subject" (*Metastases* 115). Understood
differently, we refer to this phenomenon as the "fugitive subject," a
subject escaping state machinery, fixity, moving subjunctively or transversally, possibly without willful purpose or direction.

Ferdinand's immediate response to the ontological crisis caused by
his sister's sexuality and class debasement, as the purveyor of the traumatic question "what does the woman want?" is to accumulate pros-
thetic objects that serve as agents of mediation, enabling him to
acquire distance from his crisis of Master systems or, more simply, of
"masters," the sociopolitical conductors that are part and parcel of the
state machinery. Indeed, in addition to broaching Eucharistic controversies, Webster's play is poised on the threshold that separates nascent
capitalism from feudalism, self-interest from *civitas*, invoking two dia-
metrically opposed visions of citizenship which, in turn, are presided
over by two very different authorities or "masters." If, as Žižek explains,
precapitalistic society was structured by a belief in the Master as an
abstract regulatory agent, a figure who reflects a stable paternal signifier
and exists as a model of the "ethics of self-mastery and 'just measure' "

(*Tarrying* 210), then within capitalism, this regulatory function is suspended and, rather than serving as an agent of prohibition, the Master's imperative is reconstituted as an invitation to *enjoy*. Consequently, according to Žižek, the paradox of capitalistic "society" is that when individual enjoyment is elevated to the level of "civic duty," the state machinery breaks down, and society itself ceases to exist. (Nevertheless, Žižek is wrong, as the 2004 presidential election in United States demonstrated: Bush was elected precisely because of belief in a Master for which Bush was seen as a key sociopolitical conductor; but Žižek is right to acknowledge enjoyment as the anticipated result.) Although Ferdinand initially responds to his crisis of Master referents in the ascetic or "good Catholic" mode by seeking mediation for his desire, all of the remedies he employs, namely, the poniard, mandrake root, and rhubarb, are either phallic objects or aphrodisiacs, revealing his conflation of Master systems in the extent to which his sense of duty is permeated by desire for enjoyment.

For example, at the conclusion of the brothers' explanation that the Duchess is not to remarry without their express permission, Ferdinand confronts his sister with an apparent non-sequitur, exclaiming: "You are my sister, / This was my father's poniard: do you see? / I'd be loth to see't look rusty, 'cause 'twas his" (1.1.321–3). Here Ferdinand nobly assays a compromise between duty and desire, asceticism and indulgence by urging his sister to recognize or "see" his phallic prowess while simultaneously foisting off agency on her with an appeal to masochism: he presents her with the poniard in hopes that she will use it on him, rendering him the martyr of her lust.[13] When he eventually learns that the Duchess has married and born several children, Ferdinand's response is to seek out another distancing device that is likewise penetrated by desire – a mandrake root – as he proceeds to extort from it the confession that his sister is "damned; she's loose i' th' hilts, / Grown a notorious strumpet" (2.5.3–4). Prompted by the root's professed erotic properties, Ferdinand reconciles his conflicting imperatives by imagining the sexual punishment of his sister, vicariously satisfying his lust by envisioning her "in the shameful act of sin," dominated by "some strong thighed bargeman, / Or one o' th' woodyard, that can quoit the sledge, / Or toss the bar, or else some lovely squire / That carries coals up to her privy lodgings" (2.5.42–45).[14]

It is telling that Ferdinand gets so carried away at this point that he actually inserts himself into this rape scenario. Crying for rhubarb, yet another salve, to "purge [his] choler," Ferdinand exclaims:

Would I could be one,
That I might toss her palace 'bout her ears,
Root up her goodly forests, blast her meads,
And lay her general territory as waste
As she hath done her honours.

(2.5.17–21)

What is striking about this passage is Ferdinand's methodical prescription of a hysterectomy as the "answer" to his sister's unbearable desire. Yet, what it more profoundly exposes is the hystericization of the masochist, who, Žižek explains, is suddenly "horrified at the prospect of being reduced in the eyes of the Other" to "the role of an object-instrument" and, therefore, resorts to "violence aimed at the Other" (*Metastases* 92–3). Ferdinand's shift from masochism to sadism suggests a backlash against the failure of the system of mediation posed by patriarchal absolutism working in tandem with the Catholic Church. Indeed, his later attempts, through various forms of mortification, to convert his sister into a kind of relic signal the last gasps of his Catholic faith, as he attempts to substantiate her body as a cure for his riven subjective territory. As Scott Dudley observes, the display of bodies and attenuated parts in *Duchess*, like the severed hand with which Ferdinand tortures his sister, bespeaks the play's imbrication in the Counter-Reformation obsession with bodily relics as proof of the supremacy of the Catholic Church, as tokens of saints' bodies came to be associated with miracles and cures (285–6). Ironically, Ferdinand's morbid efforts to sanctify his sister's flesh-as-spirit in order to cure his own incestuous desire fail by way of the Duchess's living parody of transubstantiation, as the trail of children she leaves in her wake testify to the miracle of her flesh. It is not surprising, then, that Ferdinand is driven to kill the Duchess; what is remarkable is how he manages to compensate for the void carved out by her death by becoming a good citizen.

Positive differentiation

Wish me good speed
For I am going into a wilderness,
Where I shall find nor path, nor friendly clew
To be my guide.

(1.1.349–52)

In exploring the impact of the Eucharistic controversies on the literature of Protestant England, Stephen Greenblatt asserts that the root of "most of the significant and sustained thinking in the period about the nature of ... figuration, ... lies less in the problem of the sign than in what I will call *the problem of the leftover*, that is, the status of the material remainder" (342). We want to challenge this assumption by arguing that the most radical dimension of Protestant thinking lies not in the material remainder but, rather, in the problem of what *is not there*: the vanishing mediator. In *Duchess*, there is no question that prior to his sister's death, Ferdinand is obsessed with precisely the "problem of the leftover," the kernel of enjoyment that resists every socio-symbolic effort to vanquish its forbidden pleasure in the interest of the Law and the official territory it substantiates. In order to understand the point at which the vanishing mediator comes into play, however, we need to go one step further to explore Webster's play as a staging of the Lacanian dictum that "there is no sexual relationship" (Žižek, *Reader* 124).[15]

It is an axiom of psychoanalysis that, in society, the sexual relationship is structurally "impossible" because the only absolute sexual relationship is incest. For society to function the incest taboo becomes Law and the primordial enjoyment of the mother, sister, or daughter is subjugated and replaced with the exogamic pursuit of "woman." Fantasy, even when subjunctively explored rather than "consciousness adrift," is an effort to organize and lend meaning to this secondary pursuit; it is what organizes our desire in socially-acceptable ways because it protects us from the incestuous urge that underlies it. The contribution of Webster's play to the debates over figuration that Greenblatt cites, then, is precisely his staging of Ferdinand's process of "going through" fantasy and, hence, *beyond* the material remainder to the dimension of "positive difference" that resides beyond the compensatory gestures of signification. This is precisely the aspect of radical Protestantism that the Eucharistic controversy seeks to mask: that beyond the signifier there may be nothing but the death drive and the incestuous implosion whereby the desiring subject becomes his own object that the subject was all along. Is this the sacrifice that the Protestant God ultimately demands of Renaissance writers from Wyatt to Donne which, once analyzed in transversal terms, is, ironically, not a sacrifice after all? The only answer to the question, "what does this inscrutable God want?" is the very same as the answer to the question "what does the woman want?": nothing less than everything. This is *not* to say, quite literally, nothing, the affirmation of "the Void that 'is' the subject," as Žižek would have it. Alternatively, following Gilles Deleuze and Félix Guattari,[16] transversal theory maintains that desire is the subject and object of desire, and not predicated on lack; the subject is instantiated by its own presence, rather than in contrast to absence or via negation, thus sharing the logic here concisely expressed by Robert Nozick:

> If something cannot be created out of nothing, then, since there is something, it didn't come from nothing. And there never was a time when there was only nothing. If ever nothing was the natural state, which obtained, then something could never have arisen. But there is something. So nothingness is not the natural state; if there is a natural state, it is somethingness. (If nothingness were the natural state, we never could have gotten to something – we couldn't have gotten here from there.) (125)

Transversal theory sees everything and, quite literally, refers to every-thing present or absent, real or imaginary, material or immaterial, and is only interested in differentiating among these terms conditionally and as circumstances, ethical or unethical, encourage doing so, depending on the particular purpose and stakes.

In order to "see" the operation of the vanishing mediator, then, we have to recognize how woman functions as the object of fantasy, from the individual woman-thing to the collective Nation-thing, and then drops out of the equation to produce the subject's encounter with radical negativity or "no thing." And we must do this, if our goal is to prevent exploitative mediation of this vanishing sort and, therefore, to offer a more empowering alternative for the construction of woman in dis-course. This alternative emerges when woman transposes into a fugitive force of presence, thereby becoming a source of transversal power by which mediation can happen without vanishing, the medium continu-ing to productively occupy the space of both its own inhabitance and the territory it works to spatialize for inhabitance by others. In *Duchess*, Ferdinand's affliction with lycanthropy renders this vanishing act visible, thereby demonstrating a process of becomings in the foreground. Moreover, it exposes how the compensations of self-help reverse the idea of subjective destitution, in "Catholic" fashion, by filling in the void and projecting the search for the werewolf within onto the perceived enemies of our obscene enjoyment, that is, our own excessive (capital-istic) way of life. In *Duchess*, Ferdinand finally arrives at a different, more honest and more ethical vision of citizenship by criminalizing *himself* and, in so doing, reconciling his clash of Master systems, enjoyment, and the Law by becoming simultaneously the vanishing mediator of this regulatory process and a subjunctive gateway for hopeful possibilities. We have hoped to show that beyond the Duchess's role as the primary vanishing mediator who precipitates Ferdinand's descent into lycanthropy in the first place, her spectral significance, her persistent presence, is what spurs comings-to-be away from state sanctioned subjectivity and facilitates possibilities for becomings-other that are revealed by Ferdinand's lycanthropic movement.

Although the exact source of werewolf syndrome in early modern culture is disputed, the cause is almost invariably attributed to an encounter with a suspected witch. The subject's affliction with werewolf syndrome is, in other words, the positivization of the missing witch, the material trace that denotes everywhere the witch is, indeed her everywhereness. Webster suggests this process of materialization not only in Ferdinand's numerous allusions to his sister's bewitching sexuality (he actually uses the word "witch" three times to describe the Duchess), but also in the fact that he only "becomes" a werewolf once she has transmuted into some other form. Indeed, when he discovers that Bosola has fulfilled his wish to see his sister murdered, Ferdinand is overwhelmed with remorse and explains that he was "distracted of [his] wits" when he bade Bosola "[g]o kill my dearest friend" (4.2.271–2). At this moment, Webster exposes how fantasy is our dearest friend and how, in "going through fantasy," we often transcend things, manifested objects that we experience as standing in the way of the fulfillment of desire; for fantasy is, to be sure, a crucial "thing" we have at our disposal: it can be the very prop and substance of our subjective territory. On this note, consider some words of David Hume for which transversal poetics is empathetic:

For my part, when I enter most intimately into what I call *myself*, I always stumble on some particular perception or other, of heat or cold, light or shade, love or hatred, pain or pleasure. I never catch *myself* at any time without a perception, and can never observe anything but the perception. (67)

Hence, in showing Ferdinand attended by a doctor as he attempts to "throttle" his own shadow (5.2.38), having explained to the others that he howls and digs up bones because he is "a wolf, only the difference / [I]s a wolf's skin [i]s hairy on the outside, / His on the inside" (5.2.16–18), Webster represents Ferdinand not as insane but as brutally lucid. For although the doctor rightly diagnoses Ferdinand with "lyc-anthropia" as an effort to explain the wolfish fiction that the Duke has come to live by, the doctor fails to recognize the truth of its content, dismissing Ferdinand's ravings as nonsense. When Ferdinand tells the doctor that "Physicians are like kings, / They brook no contradiction" (5.2.66–7), the doctor completely misses the point that Ferdinand's pursuit of werewolf syndrome-as-"self-help" reveals: that he cannot but be true to himself in becoming other than what he was, by embodying the contradiction that makes death – that is, the death of the Lacanian fantasy of the subject-as-void – the greatest act of self-love. In other words, without his sister as the scapegoat of his impossible desire, Ferdinand lacks the recourse of fantasy, and he is forced into a far more terrifying acknowledgment of the "no thing" that is the something, paradoxically, his very being at the moment of his death: "The pain's nothing; pain many times is taken away with the apprehension of greater, as the tooth ache with the sight of a barber that comes to pull it out: there's philosophy for you" (5.5.58–61). Ferdinand dies as his own barber, as the only one who can cut the hair that grows inside him. He dies, therefore, a good citizen, himself the vanishing mediator, and not woman, of an inestimable truth about the subjective wilderness we may all contend with on the inside.

Herein lays the paradox of self-help. We enter into the "wilderness" of an identity crisis assured that we can change inasmuch as we accept the pain of change only to the extent that, underneath all the props, we fantasize that we can be who we really are, as Robbins suggests when he explains that no force is greater than "the drive to preserve the integrity of our own identity" (127). Put differently, the "miracle" of self-help, the outcome of the steady process of accretion that expands our "Master system" of referents, is the miracle of fantasy, the belief that our identity is something that lies outside of us, in the vast desert of our sacred fetishes, whereas identity actually rides on the ebb and flow of our becomings and comings-to-be, as our subjective territory spatializes as it is spatialized in negotiation with state machinery and fugitive forces of transversal power by which it reconfigures, expands, and transforms. Externalization, then, is not opposed to implosion of the self, what Žižek, following Adorno, understands as "the 'post-psychological' individual of the 'totalitarian' society" (*Metastases* 21; *Gaze and Voice* 208–49), but is rather symptomatic of the positive differences among the self-as-master, the self-as-other, and the self-as-monster that create unlimited potential to become what you are not. This may be why there are no movies about John Webster, but only affective traces of him in the Shakespeare industry – in the Shakespace to which it contributes: Webster has become a vanishing mediator for a self-help ethos that is the product of totalitarian fantasies, the interstices of which speak to the ongoing reality of werewolf syndrome.

Notes

1. Self-help was not foreign to people in early modern England, but it usually took the form of manuals regarding how to be a good Christian, mother or gentlemen.
2. For detailed analysis of love in *Shakespeare in Love*, see Donald Hedrick and Bryan Reynolds, "Shakespace and Transversal Power," in their edited volume, *Shakespeare Without Class: Misappropriations of Cultural Capital* (New York: Palgrave Macmillan, 2000).
3. See Bryan Reynolds, "The Devil's House, 'or worse': Transversal Power and Antitheatrical Discourse in Early Modern England" (*Theatre Journal* 49.2 [1997], 143–67); *Becoming Criminal: Transversal Performance and Cultural Dissidence in Early Modern England* (Baltimore, 2002), 1–22; and *Performing Transversally: Reimagining Shakespeare and the Critical Future* (New York, 2003), 1–28.
4. Fredric Jameson, "The Vanishing Mediator; or, Max Weber as Storyteller," *The Ideologies of Theory*, 2 vols. (Minneapolis: University of Minnesota Press, 1988), vol. 2: 3-34 (originally published in *New German Critique* 1 [1973]).
5. For example, see Slavoj Žižek, *For They Know Not What They Do: Enjoyment as a Political Factor* (London: Verso, 1991), 90.
6. On "objective agency," see Reynolds, *Becoming Criminal*: 24–7.
7. See Lynn Enterline, " 'Hairy on the In-Side': *The Duchess of Malfi* and the Body of Lycanthropy," in *The Yale Journal of Criticism: Interpretation in the Humanities*. 7.2 (Fall 1994): 85–129.
8. For an intriguing psychoanalytic reading of Ferdinand's melancholia in relation to the Duchess's sexuality, see Lynn Enterline, *The Tears of Narcissus: Melancholia and Masculinity in Early Modern Writing* (Stanford: Stanford University Press, 1995). 242–303.
9. In addition to Chapters 1,3,8, and 10 in this book, see Reynolds, "The Devil's House, 'or worse' ": 143–67; *Becoming Criminal*: 1–22; and *Performing Transversally*: 1–28.
10. See Karen S. Coddon, "*The Duchess of Malfi*: Tyranny and Spectacle in Jacobean Drama," in Dympna Callaghan ed., *The Duchess of Malfi* (New York: St. Martin's Press, 2000): 25–45.
11. See Frank Whigham, "Incest and Ideology: *The Duchess of Malfi* (1614)," in David Scott Kastan and Peter Stallybrass eds., *Staging the Renaissance: Reinterpretations of Elizabethan and Jacobean Drama* (New York: Routledge, 1991): 263–74.
12. On Ferdinand's sexual desire, see Kathleen McLuskie, "Drama and Sexual Politics: the Case of Webster's Duchess," in Callaghan ed., *The Duchess of Malfi*: 104–21.

13. For a potential alternative reading of this scene, consider Bryan Reynolds and Janna Segal's discussion of Desdemona's masochism in Joseph Fitzpatrick & Bryan Reynolds (with additional dialogue by Bryan Reynolds & Janna Segal), "Venetian Ideology or Transversal Power? Iago's Motives and the Means by which Othello Falls," in Reynolds, *Performing Transversally*: 53–84.
14. See Whigham, "Incest and Ideology," for an astute reading of these lines regarding class issues.
15. See also Žižek, *The Sublime Object of Ideology* (London: Verso, 1989): 126.
16. See Gilles Deleuze and Félix Guattari, *Anti-Oedipus: Capitalism and Schizophrenia*, trans. Robert Hurley, Mark Seem and Helen R. Lane (Minneapolis: University of Minnesota Press, 1983), 25.

Works cited

Adorno, Theodor. *Freudian Theory and the Pattern of Fascist Propaganda*. London: Routledge, 1991.

Cooke, Deryck. *I Saw the World End: a Study of Wagner's Ring*. London; New York: Oxford University Press, 1979.

Dudley, Scott. "Conferring with the Dead: Necrophilia and Nostalgia in the Seventeenth Century." *ELH*. 66.2 (Summer 1999): 277–94.

Greenblatt, Stephen. "Remnants of the Sacred in Early Modern England." *Subject and Object in Renaissance Culture*. Ed. Margreta de Grazia, Maureen Quilligan, and Peter Stallybrass. Cambridge: Cambridge University Press, 1996. 337–45.

Hume, David. *Treatise of Human Nature*. Oxford: Clarendon Press, 1967.

Nozick, Robert. *Philosophical Explanations*. Cambridge: Harvard University Press, 1981.

Robbins, Anthony. *Awaken the Giant Within: How to Take Immediate Control of Your Mental, Emotional, Physical, and Financial Destiny*. New York: Simon & Schuster, 1992.

Saleci, Renata and Slavoj Žižek Eds. *Gaze and Voice as Love Objects*. Durham and London: Duke University Press, 1996.

Žižek, Slavoj. *Tarrying with the Negative: Kant, Hegel, and the Critique of Ideology*. Durham, NC: Duke University Press, 1993.

—— *The Metastases Of Enjoyment: Six Essays On Woman And Causality (Wo Es War)*. London and New York: Verso, 1994.

—— "There is Sexual Relationship." Ed. Renata Saleci, and Slavoj Žižek. *Gaze and Voice as Love Objects*. Durham and London: Duke University Press, 1996. 208–49.

—— *The Žižek Reader*. Eds. Elizabeth Wright and Edmond Wright. London: Blackwell, 1999.

—— *Did Somebody Say Totalitarianism? Five Essays in the (Mis)Use of a Notion*. London and New York: Verso, 2001.

Webster, John. *The Duchess of Malfi*. *English Renaissance Drama: a Norton Anthology*. Eds. Katherine Eisaman Maus and David Bevington. New York: Norton, 2002. 1755–830.

—— *The White Devil*. *English Renaissance Drama: a Norton Anthology*. Eds. Katherine Eisaman Maus and David Bevington. New York: Norton, 2002. 1664–748.

10

Performative Transversations: Collaborations Through and Beyond Greene's *Friar Bacon and Friar Bungay*

Bryan Reynolds & Henry Turner

> Beginning in sixteenth-century England, a distinct criminal culture of rogues, vagabonds, gypsies, beggars, cony-catchers, cutpurses, and prostitutes emerged and flourished. This community was self-defined by the criminal conduct and dissident thought promoted by its members, and officially defined by and against the dominant preconceptions of English cultural normality. In this book I argue that this amalgamated criminal culture, consisting of a diverse population with much racial, ethnic, and etiological ambiguity, was united by its own aesthetic, ideology, language, and lifestyle. In effect, this criminal culture constituted a subnation that illegitimately occupied material and conceptual space within the English nation. With its own laws and customs, it was both independent of and dependent on England's official (mainstream) culture. It was self-governing but needed the law-abiding populace for food and shelter and as a social entity against which to define itself. I also argue that the enduring presence of this criminal culture markedly affected the official culture's aesthetic sensibilities, systems of belief, and socioeconomic organization. It was both conducted by and a conductor for what I call "transversal power." (Bryan Reynolds, *Becoming Criminal* 1)

Beginning in sixteenth-century England, a distinct academic culture of friars, professors, mathematicians, magicians, astrologers, adepts, and students emerged and flourished. This community was self-defined by the academic conduct and dissident thought promoted by its members, and officially defined by and against the dominant preconceptions of English cultural normality. In this concluding chapter we argue that this amalgamated academic culture, consisting of a diverse population with much etiological ambiguity, was united by its own aesthetic, ideology, language, and lifestyle. In effect, this academic culture constituted a subnation that illegitimately occupied material and conceptual space within the English nation. With its own laws and customs, it was both independent of and dependent on England's official (mainstream) culture. It was self-governing but needed the law-abiding populace for food and shelter and as a social entity against which to define itself. We also argue that the enduring presence of this academic culture markedly affected the official culture's aesthetic sensibilities, systems of belief, and socioeconomic organization. It was both conducted by and a conductor for what we call "transversal power."

The repetition, replacement, and commonality effected between the preceding two paragraphs – the correlation between one form of heterodoxy and another, situated historically and effected by a substitution of terms and positions – marks an opening in a broader set of arguments concerning the genealogy of modern *"homo academicus."* These arguments, which at their core pertain to the writing, now, of this very book, pertain also to the relationship among the official culture of the university and the relative force of academic discourse. This force may be either transversal or state-oriented – a force that works either to effect conceptual, emotional, and/or material flux or that works to consolidate stability. Hence our analysis has several purposes, some historical, having to do with the once past-present now absent-spaces of early modern England, and others pertaining to both the present-spaces and the future-present-spaces created by fugitive explorations in scholarly research. In the first place, negotiating the past-present with the absent, we will examine how Robert Greene's *Friar Bacon and Friar Bungay* raises a series of questions about the legitimating of knowledge-claims and methods of intellectual inquiry during the early modern period, about the status of so-called "scientific" thought in early modern culture, and about the relationship between the specialized and necessarily exclusive epistemologies of the academy and the more widely disseminated and normative discourses that constitute what is conventionally referred to as "popular culture." Secondly, we hope to demonstrate how these problems must be understood as extending into the beginning of the twenty-first century, when many of the structures and relationships visible in Greene's play have assumed an even more elaborate form: Friar Bacon and Friar Bungay are finally shadowy figures for ourselves, and so the arguments that follow should be read as an examination of the articulatory space that we refer to as "performative transversations" in modern academic discourse.

The historical premises of our investigation are clear enough and may be stated concisely. For specific reasons associated with the emergence of "humanism" in England and its effect on the institutional organization of the universities and their curricular emphases; with a rapidly centralizing and expansionist Tudor state; with growing interest across all levels of English society in technology, applied mathematics, and proto-"experimental" methods of inquiry; and with the development of an urban consumer culture capable of supporting an increasingly differentiated market in books and public entertainment, the position of *homo academicus*, as we shall designate him – the Bachelor and Master of Arts; the lecturer, instructor, and professor; the practitioner, secretary, or "reader"; the poet, playwright, and "man of letters" – gradually came to require a more precise and more explicit definition by early modern contemporaries. This pressure derived from primarily three sources, one "internal" and two "external" to the aspiring *homo academicus* per se: first, from the subjective territory of the academic himself, who sought to secure the prestige and influence that might attend on perceived authority and expertise in a particular field and official territory; second, from sociopolitical conductors – individuals and institutions – which disposed of wealth and power, and needed to evaluate the claims of those who sought their patronage; and third, from sociopolitical conductors – unofficial as well as official – which sought to circumscribe, negate, or otherwise discredit the authority of the academic subject, either for reasons of professional rivalry or of ideological difference. The most famous examples of this phenomenon indicate how the struggle to define the position of *homo academicus* took place well beyond the university itself as a specific field of academic practice and indeed depended crucially on the patronage of sources of power that were non-academic, as the cases of Gabriel Harvey, John Dee, Thomas Blundeville, Thomas Harriott, and many others demonstrate. This was particularly true of those men who were seeking to attain patronage through their expertise in the emerging yet nevertheless still fugitive fields of mathematics and technology.

In the character of Friar Bacon, Greene has created a figure who condenses several variables that were converging to define the position of the academic subject *inside* the university system at the end of the sixteenth century in England, and it is this position within the university field, as well as its relationship to extra-academic institutionalized power within and outside of the demarcated arenas monitored by the state machinery, that we will be considering here. Elsewhere, in our essay, "From *Homo Academicus* to *Poeta Publicus*: Celebrity and Transversal Knowledge in Robert Greene's *Friar Bacon and Friar Bungay*

(c. 1589)," we explain the relevance of Friar Bacon's position within the academic field to Greene's *own* position as *homo academicus*-becomings-*poeta publicus*: to Greene's attempt, in other words, to *distance* himself from the university in order to achieve a position in the emerging market for playwright and commercial publication.[1] In all three cases – that of Friar Bacon and Friar Bungay, that of Greene himself, and that of ourselves – the act of public performance becomes a critical point of refraction where the field of *homo academicus* may be modeled and its relationship to various sociopolitical conductors and other forms of institutionalized power understood.

Greene's play offers a remarkably sharp definition of the different fields of power in which late-sixteenth-century *homo academicus* found himself positioned: as a discrete institutional entity, the University of Oxford is imagined as an official territory with its own internal hier-archies, resources, and semi-autonomous identity, as indicated by the repeated references to the "academic state" (Greene 2.165) by various agents distributed throughout the sociopolitical fields represented in the play. This community is constituted through its relationship to monarchical power and the national community; through its relations with an international network of scholars and rival institutions; and through its internal separation into colleges and academic sub-communities organized around specific intellectual problems and methods of research – as is the case within the university system today. Much of the play's action may be described as a symbolic attempt to examine the relationships within and among these three primary fields as they are submitted to the stresses of contestation when phenomena of assimilation, transformation, and expulsion are at work – when what Glenn Odom and Bryan Reynolds describe as "pressurized belongings" occur (see Chapter 8), all of which typically accompany struggles for dominance:

Pressurized belongings ... are the related and often conflicting processes of assimilation and expulsion by which one becomes a member of an alternative group, subjective territory, or official territory at the expense of one or more of its members. The new member causes overflow or reconfiguration such that not all of the extant members can remain the same or remain at all if the system is to maintain equilibrium. This is not to say that the substance of the group necessarily drastically changes or that there is transforma-tion of the parameters within which the group maneuvers, but rather that there is only room for so many members of certain kinds.

Accordingly, the international rival to Friar Bacon and Friar Bungay, Jacques Vandermast, arrives in the company of the Emperor and the English King in order to test the reputation of the English institutions; his function is to assist the Emperor in the political domination of the English nation through a battle of wits and necromantic surplus – a steroid occult infusion – that takes the place of an actual military exercise, and his defeat by Friar Bacon marks a temporary alliance between two intra-national sociopolitical entities (University and Crown) in order to reassert the strength of the "native" political body and official territory. This struggle, however, also reveals the deep independence and even the antagonism between Crown and University, and it is prosecuted by Friar Bacon for several reasons: first, to secure a transfer of both material and symbolic resources from the former to the latter in exchange for a temporary political alignment and the expenditure of the university's symbolic capital; and, secondly, and perhaps more importantly, in order to secure for Bacon himself the symbolic capital that derives from royal recognition and gratitude in order to buttress his own position and intensify his affective presence – the combined material, symbolic, and imaginary existence of a concept/object/subject/event and its multiplicities – within the strained economy of the university field, where his long-standing research program has suddenly been discredited.

Such are the broad outlines that structure the play's imaginative fiction, which, through the principle of homology may be seen to correspond to the structure of the international and national political field at the end of the sixteenth century in the West. We have taken the principle of homology from the work of Pierre Bourdieu,[2] where the term designates an analogical similarity in structural situation between different subfields in a society: therefore, a dominant position in the economic field may often be correlated with a dominant position in the field of political and legal power, while a dominated position in one field often corresponds with a dominated position in another, thereby stratifying subjective and official territories structurally and methodologically, if not also sociopolitically and ideologically. As we argue in "From *Homo Academicus* to *Poeta Publicus*," the position of the playwright in England at the end of the sixteenth century provides a particularly strong example of the principle of homology and the complexity of cultural analysis that it permits: the playwright is in a relatively dominated position both economically and politically, and this homological correspondence itself enables a separate homological identification with other dominated positions and the expression of this identification in symbolic form. At the same time, as Bourdieu argues, the playwright's real social power lies in his capacity to nominate meaningful sociopolitical categories and to control, in part, their representation and circulation in symbolic form; and in this he enjoys a dominant position that is out of proportion to – one that tends to invert – his economic and legal power.

Hence, the principle of homology provides an important methodological tool that may account for the heterodox, fugitive, or transversal attitudes often visible in early modern plays, especially when applied in conjunction with what we call the "principle of citationality," which refers to a layering of individually legible codes – and therefore often also laminated ideologies – represented on stage, in any social performance, or in any expressive medium, such as this one. Through their simultaneous yet staggered articulation, citational codes concentrate and extend the significance of the-code-that-can-be-cited (the very definition of a code), but at the same time refract that code, enabling a defamiliarizing and analytic gaze that the early moderns called "theater" – that ancient technology of beholding – and that we call "theory." Under the effects of homology and citationality – double-fisted theoretical punches that strike at the heart of legitimate sources of power and belief – the dominated position of the playwright vis à vis the early modern market, state machinery, and official culture suddenly shudders into view, exposing at the same time how heterodox and fugitive attitudes can intercede and/or emerge from dominant methodological traditions within academic discourse by means of performative transversations: the invention of new articulatory spaces, new theoretical languages, and new speaking voices for a more challenging academic discourse.

The very language of Greene's *Friar Bacon and Friar Bungay* reveals its most important homological correspondence, one that still applies to certain sociopolitical conductors within state machineries today that constitute and preside over official avenues for academia and the arts: the position of Friar Bacon, whom we shall now call "B", within the "academic state" (2.165) finds its structural analogue in the position of the English King, whom we shall now refer to as simply "Henry," within the international political field. Because of this homological correspondence, the resolution of the different levels of political conflict require a convergence between these two figures and their symmetrical positioning – at the end of the play – at the apex of their respective fields of power. Consequently, England's position in relation to other international states is structurally analogous to other scales of relationship, such as the position of the University of Oxford in relation to the national political field of England and Brazennose College's position in relation to the University "state" at large; thus B's relationship to other academic potentates may be directly correlated with that of Henry in relation to the Emperor who visits Oxford with him. This homology finds expression in the defensive analogies that both Henry and B employ: England, Henry declares, is "ringed with the walls of old Oceanus" (4.2), much the way B proposes to "circle England round with brass" (2.172); the rich intellectual capital of the schools is described in terms of royal forests and landscapes, "fat and fallow deer" (9.4), and so forth.

But it is important to emphasize that, as a heuristic category, the principle of homology demonstrates not only correspondence, but also the tension, friction, and contestation that inevitably exist between social fields with different and often conflicting currencies of value; indeed, we may suspect that as homological alignments become increasingly triumphant in their representation, so also increases the degree of conflict, according to the principle of pressurized belongings, that potentially resides in their conjunction. Whatever the extravagance of Henry's rhetoric, B's research program in fact belies Henry's claims for England's natural sovereignty. As the Armada crisis had recently showed, the protection afforded by the seas could no longer be taken for granted, and a mysterious power, imagined in theological, magical, or technological terms, had to be invented as its supplement. The breath of God blowing to founder the Spanish navy, a wall of brass compassing all of England: both are early modern avatars of late-modern fears, and of the settlement bulwarks and missile shields designed to dispel them. For B to be powerful, in short, England must appear to be in a state of insufficiency, and for this reason, Henry needs to *expose* B to the threat of Vandermast as much as he needs him to vanquish that threat, if only to remind him of the benefits of royal protection. For the same reason, he needs to supervise their disputation as an authorizing witness, crucially performing the "witness-function" in the hermeneutic equation (see Chapter 7), to the contest and as the guarantor of B's enduring reputation and affective presence. By doing so, furthermore, Henry manages to accomplish a contradictory goal with a single dialectical gesture, marginalizing B in the very moment that he celebrates him. By commanding *homo academicus* to demonstrate his ability, Henry transforms years of scholarship and research into a momentary, spectacular performance whose power is symbolic rather than material: not a physical transformation of substance or the erection of a technological marvel, but arcane words and arguments that no one but B and Vandermast, locked in a titanic struggle for their own anachronism, understand.

But – by virtue of homology and citationality – is not "Vandermast" simply another Henry in this scene, which reflects the common critique of academic discourse today, particularly when enhanced by the language of theory? It is a critique that often rears its head within the very system of which it is mutually symptomatic; and so the scene becomes its own delightful theater, staging the becomings and comings-to-be of *homo academicus* in relation to the communities he needs to both engender and defy in order to survive and replicate. In the play (but is this just a play we are discussing? for B is certainly in the theater, yet he is also in the text), B himself is under attack – and more vulnerably so because he does not have the support of a community of which he is a devoted member – not simply by Vandermast but by rivals from within his own field, much the way Henry's authority is hedged not simply by foreign rivals but by competing intra-national sociopolitical conductors of such institutions as the University and the Church; in this field, too, the very singularity that makes B so powerful – the only and last defense of both England and the "academic state" (2.165) – is what makes him threatening to the hierarchy within the university community as well as to Henry; his transversality, unique and affective, cannot be effectively contained or channeled. The University and the Church, too, attend the disputation and have no less of an investment in the resolution of "the doubtful question," as Vandermast describes it (9.23). Here the stakes are nothing less than the authorization of an entire research program: a set of problems, techniques, and goals; a specialized vocabulary; a canon of textual authorities. The entire play, in fact, provides a perfect demonstration of the attempt to formulate a scientific paradigm, in Thomas Kuhn's terms, and to define the terms in which the legitimacy of that paradigm might be recognized and its symbolic power secured. The real importance of the "Brazen Head" resides in its symbolic status as a catalyst for dissective-cohesive aspirations that must give way to investigative-expansive processes: in the epigrammatic formula of its pronouncements, signifiers of a larger and more mysterious potential power to transform all things; in its status as an exemplary construction, a model project that B may wield in order to legitimate his theoretical paradigm, demonstrate competence, promise success in future endeavors, and receive material support; and, not least, in its metonymical name, the "Brazen Head" that will secure the reputation of "Brazennose College" within the university, national, and international communities, and secure B's place at the center – as a key sociopolitical conductor with extraordinary "emulative authority" (see Chapter 1) – of the cascading homological series.

In this respect, the Brazen Head is simply a spectacular manifestation of B's occult powers, of expanding subjective and official territories, and thus of the changing status of mathematics, technology, and the mechanical arts that was taking place in early modern England during the final third of the sixteenth century, both inside the university and outside it, largely through the efforts of "inventive" academic subjects like John Dee, Gabriel Harvey, Blundeville, Harriott, Sir Henry Savile, and many others. As a result, Dee, in his famous "Mathematical Praeface" to Henry Billingsley's English translation of Euclid's *Elements* (1570), could claim for "*mathesis*" a variety of philosophical, theological, technological, and occult properties, much the way B's "magic" and "mathematic rules" exist in an expanding rhizome of knowledge-practices that include astronomy, navigation ("tides and ebbs"), necromancy, pyromancy, and aeromancy. Through a transversal act of historical imagination, Greene has retrojected contemporary developments in mathematics into a romanticized, medieval past, drawing on the longstanding tradition of proto-scientific inquiry at Oxford as a way of legitimizing his own *alma mater* while also preserving a certain skeptical and amused distance. The Brazen Head, after all, finally proves useless, a ridiculous stage effect, and in this way Greene is, like Henry, asserting the superiority of his own specialized language – poetics and dramatic representation – over the spurious incantations of B's device.[3]

In doing so, however, Greene (like B) also dramatizes a contradiction that is fundamental to all academic research and to every academic utterance, which results from its position at the point of intersection between two competing fields. To the power of the King, official culture, and state machinery – of any force that works in the interest of coherence – the academic utterance must retain a reserve of authority that can fund this power's initiatives, even as these remain entirely distinct from, and often in direct contradiction to, the academic research necessary to produce further authoritative utterances. State power, in short, must cultivate "science" for rhetorical purposes even as it circumscribes it with separate gestures calculated to undercut it and ensure its final impotence and irrelevance outside its own immediate field. To this end, state power inflates the currency of academic prestige within the "official" territory of academe and fosters vigorous competition over the few resources that it allocates, in this way ensuring that the currency of any stature within the academic field will be achieved at enormous personal and material cost, restricted to an extreme minority, and remain difficult to convert into any form of power, symbolic, transversal, or otherwise, outside of the academic field that defines it.

This inhibiting dynamic – state power's ability to "divide and conquer" – explains why B can occupy such a marginalized position in the academic system and yet play such a central and singular symbolic role in the defense of the academic field; why he occupies a poor single cell, employs a foolish research assistant, eats meager meals, embarks on exhausting and ultimately fruitless projects, and yet still manages to espouse a secret, specialized form of knowledge that fascinates and provokes competitive rivalry and anxiety from his colleagues and garners extravagant praise, but only praise, from Henry at the end of the play. In the play, recognition is indeed the primary currency of the "academic state" (2.165), and remains the only meal ticket of *homo academicus*; B, after all, needs the recognition of his colleagues as much as they, as a collective body, need the national and international recognition that his triumph over Vandermast before Henry and foreign potentates brings. The purpose of Friar Bungay, B's inferior double-*cum*-collaborator (again, another Henry in this scene), is to represent in a particularly compressed way the intra-collegial relationships that are the consequence of the larger relationship between the academic field and the complex social network over which state power strives to reign. As exemplified by B and Bungay's relationship, "collegiality" consists in a competitive dependency structured by the threat of humiliation, a peculiar combination of solicitude and secrecy that characterizes conversations among near-equals, in disingenuous declarations of solidarity and gratitude, all in the interest of achieving the affective presence and emulative authority requisite for the maintenance of position. It is a world of pressurized belongings that notable university men such as Harvey, Greene, and Nashe must have known all-too-well; for all three it provided a resource of spleen that spilled over into their public satires and may even have goaded all three, to different degrees, to a seek an alternative position outside of the academic field in the market of commercial publication. Had the three men engaged in performative transversations, becoming an emergent community within the academic and/or commercial fields, as many of their contemporaries did, then they may have been more effective at fostering *homo academicus* – not in its fantasized singularity, but rather as a member of a community of friars, professors, mathematicians, magicians, astrologers, adepts, and students who would more appropriately find homology in the theater. As figures for both B and Henry, we are pointing toward an academic culture that sees performance as its first principle – the research, formulation, dialogue, expression, and teaching of ideas and art – on which all collaboration depends and to which academic discourse should aspire.

Notes

1. See Reynolds and Turner, "From *Homo Academicus* to *Poeta Publicus*: Celebrity and Transversal Knowledge in Robert Greene's *Friar Bacon and Friar Bungay* (c. 1589)," in Edward Gieskes and Kirk Melnikoff eds., *Writing Robert Greene: New Essays on England's First Professional Writer* (New York: Ashgate, forthcoming 2006.).
2. Pierre Bourdieu, "The Field of Cultural Production, or: The Economic World Reversed," trans. Richard Nice in *The Field of Cultural Production: Essays on Art and Literature*, ed. and introduction by Randal Johnson (New York: Columbia University Press, 1993), 29–73.
3. On the homologies between Greene and Bacon, playwright and *homo academicus*, see Reynolds and Turner, "From *Homo Academicus* to *Poeta Publicus*: Celebrity and Transversal Knowledge in Robert Greene's *Friar Bacon and Friar Bungay* (c. 1589)."

Works cited

Greene, Robert. *Friar Bacon and Friar Bungay. English Renaissance Drama: a Norton Anthology*. Eds. Katherine Eisaman Maus and David Bevington. New York: Norton, 2002. 134–80.

Reynolds, Bryan. *Becoming Criminal: Transversal Performance and Cultural Dissidence in Early Modern England*. Baltimore: Johns Hopkins University Press, 2002.

Afterword: Re: connaissance

Bruce R. Smith

Reader, you've been conned.

In the first place, by theater. Early modern actors were infamous in their own time as con-men. Their stock in trade was deception: they pretended to be people they were not. Sir Francis Bacon speaks for many of his contemporaries in condemning the sleights of hand, voice, and genitals that turned commoners into nobles and boys into women: "You shall now see them on the stage play a king, or an emperor, or a duke; but they are no sooner off the stage but they are base rascals, vagabond abjects, and porterly hirelings, which is their natural and original condition" (qtd. in Ashley 171). In the eyes of civic authorities, con-men actors attracted con-men criminals who plied their trades among the spectators while the actors plied their trade on the stage. In a petition dated 28 July 1597 the Lord Mayor and aldermen of the City of London asked the queen's Privy Council to intercede and stop stage plays in the suburbs that lay beyond the city's jurisdiction. Among the several "inconveniences" caused by plays the mayor and aldermen cited the fact that playhouses "are the ordinary places for vagrant persons, masterless men, thieves, horse-stealers, whoremongers, coz-eners, coney-catchers, contrivers of treason and other idle and danger-ous persons to meet together and to make their matches to the great displeasure of Almighty God and the hurt and annoyance of her Majesty's people" (qtd. in Ashley 167). Cozeners and coney-catchers, if not whoremongers and treasonous rabble-rousers, depended, like the actors, on securing the complicity of their victims. In *Transversal Enter-prises* Bryan Reynolds and his collaborators have seized on this early modern social anxiety and made it the groundwork for a postmodern critical enterprise. "The transversal influence of the theater," Reynolds says in Chapter 1, "was most manifest in the workings of criminals and social deviants, such as individuals who disguised themselves as gypsies in order to extort money, sell herbal remedies, and read palms; con-men who pretended to be different people in order to perpetrate crimes; and people who practiced transvestism, whether male-to-female in the theater or female-to-male on London's streets" (13 in this book). The con-man/contra-man and his "mark"/victim come to particular prominence in Chapter 3, when Reynolds and Anthony Kubiak find in Hamlet a startling instance of "the con-man/contra-man, who remains largely unconcerned with the mark's recognition of being-conned. It is of no consequence to the con-man whether the mark ever finds out. It is the game that gives pleasure – the play." If Hamlet is the con-man, then the audience – that's you and me – are his mark: "We are, Hamlet (or Shakespeare, or Jacobi, or someone ...) tells us to our faces, fools" (72 in this book).

The "con" in con-man is "confidence": it involves placing trust, reliance, or faith in a person or a thing.[1] (Take Paulina, for example, when she bids Hermione's statue come to life: "it is required / You do awake your faith." In the next breath she worries aloud that someone will think "it is unlawful business / I am about" [*The Winter's Tale* 5.3.94–7]). Etymologically, confidence is "with+faith." It is that quality of granting truth-value to pretense that makes theater, in the eyes of Reynolds and the others who share these pages, a "subjunctive space," a place where "as if ... " and "what if ... " are the governing propositions. "Paused consciousness" is another descriptor that Reynolds et al. say encourages this phenomenon. The theater may thus be a deceitful space, but it is a space that spectators enter in good faith. In the eyes of most poststructuralist critics, what Paulina is about is indeed "unlawful business." Spectators in 1611 may have awaked their faith, but critics trained in cultural materialism, new historicism, deconstruction, and Lacanian psychoanalytical theory make it an article of faith *not* to give their imaginative assent to theatrical illusions. The truths of economic conditions, social process, *différance*, and the Law of the Father demand skepticism, objective distance, critical intervention, and commitment to the positive truths that such reading strategies produce. The writers in this book dare to doubt that positivism. "Critical theory," as Reynolds and Kubiak observe, "especially in its more political modalities, still seems to believe itself capable of discovering the secret ideologies of capitalism, the substrate of sexism, or the representational malaise of gender itself, while believing its own agendas to be relatively transparent. Theater ... or at least theater that is the very embodiment of the Machiavellian mind, is not so sure" (81 in this book).

In the second place, then, you have also been conned, oh reader, by academic culture. Truth is not so fixed; like the progress of a play, it keeps moving. Transformation, in the vocabulary of Reynolds and his collaborators, possesses no mood: it is not transitive, not reflexive, but an intransitive state-of-becoming. "Becomings," as Reynolds defines it at the outset, "are desiring processes by which people transform into something different" (2 in this book). Notice the intransitive mood of this usually transitive, usually passive-voice verb. Reynolds and his collaborators cast themselves as "fugitive explorers." At first blush that might seem the perfect moniker for deconstruction-minded critics. But what Reynolds and company are attempting here is craftier than that. In *Transversal Enterprises* and in his earlier work Reynolds has diagnosed a major crisis in contemporary theory and devised a way through and beyond that dilemma. The crisis has two aspects. The first arises from Marxism and cultural materialism: if individual consciousness is socially determined, then what room is there for individual agency, for resistance, for social change? The second arises from deconstruction: if all meanings are the arbitrary result of marking binaries, of distinguishing *this* from *that*, then where is there ground to stand on, a place from which one can affirm meaning and proceed to act on that meaning? Reynolds has responded to these challenges by putting together his own critical methodology, "transversal poetics," and trying it out on a variety of topics, in a variety of venues, in a variety of discursive and performative genres.

Transversal theory is Reynolds' attempt to stake out and explore that "territory" or "space" of the possible, of realities alternative to those set in place by dominant institutions of religion, law, nation, and so forth. Reynolds and his collaborators carry out underground reconnaissance work. The geographical metaphor, as Reynolds notes in *Becoming Criminal*, is implicit in the root meaning of the word *transversal* as "something lying athwart" or a "deviation" (18). In Chapter 1 of this book Reynolds insists on the difference of his method from *différance:* "Unlike deconstruction, transversal poetics asks that we consider artifacts positively and extensively, rather than define negatively, defer continuously, or dismiss alternative interpretations and applications by relying only on dialectical argumentation ... fugitive inquiry, working to reveal portholes, expand passages, and foster travel into disparate territories, does not pursue or resolve comfortably with the nihilism of deconstruction or with the notion of infinity that it invokes" (9–10 in this book). Hence "theaterspace," "Shakespace," "Mary/Mollyspace," "Mary/Marlowespace," "Hamletspace," and the other territories claimed for these "transversal enterprises." Thomas Middleton and Thomas Dekker's play *The Roaring Girl*, with its confusions of male and female, criminal and hero, fiction and reality (the play's fictional subject was a real person who once appeared onstage in her own right), provide Reynolds and Janna Segal in Chapter 2 with a particularly varied stretch of transversal territory, a terrain they define as "a chaotic, boundless, challenging, and transformative space through which people traverse when they violate the conceptual and/or emotional boundaries of their prescribed subjective addresses" (37 in this book). In the course of these transversal enterprises Reynolds and his collaborators provide critiques of Althusser, Derrida, Butler, and others. Above all, they provide optimism in the face of deconstructionist nihilism. "When the deconstructionists' rhetorical dust settles discursively or continues discursively to cloud the air," Reynolds and Segal insist, "... a phenomenon remains: a human experiencing idiosyncratically a field of consciousness and possessing degrees of agency and potential peers on" (41 in this book). "Peering on" figures as a perfect pun for the collaborative quality of these enterprises.

Finally, reader, you have been conned by this book. To wit:

con-	ceivable	com-	bine	co-	gency
	cept(ual)		bobulate		gnitive
	clusion		mensurate		herence
	ditional		mercial		ordinate
	ducive		mit		(l)laborator
	ductor		mon(ality)		(l)lective
	fuse		munity		(l)lege
	junction		parable		(l)loquial
	science		parison		rollary
	scious(ness)		pass		(r)rect
	sequent		pensation		
	sequence		pete		
	sider		placency		
	spirator		plex		
	strain		ponent		
	struction		prehension		
	tain		prise		
	template				
	temporary				
	textual(ize)				
	tinence				

tingent
tortion
tribute
trol
troversial
verse
vert
vince

And that just in Chapter 1. The critical terms "transversal" and "transversality," as Reynolds notes in *Becoming Criminal*, were first used by Félix Guattari to refer to the phenomenon of *group* desire, particularly desire that takes account of other people (17).[2] It is altogether appropriate, therefore, that many of Reynolds' projects have been collaborations. *Performing Transversally* is a distinct oddity: not a single-authored book nor yet an edited anthology of critical essays by several hands but a book that includes seven of ten chapters that were written in collaboration with others (sometimes two others), as well as a foreword and afterword by scholars who offer their views of a manuscript they encountered only when it was otherwise complete. A statement in the first chapter of *Performing Transversally* explains why this should be the case: "Transversal theory ... encourages the getting outside of one's subjective territory synchronically, diachronically, and fantastically in order to interact and merge with the subjective territories of others, pushing the limits of emotional and conceptual ranges" (41).

The book you have in hand continues that spirit of collaboration. Or is the apter word "collusion"? Not a laboring together but a playing (*ludere*) together of many people's imaginations. I find myself imaginatively engaged by the attempt here to enact the very process of coming-to-be that constitutes this book's subject. The result, for sympathetic readers, offers a new form of *consciousness*, literally a "with+knowing," a knowing together. With what? With whom? When? Where? How? In the first instance, consciousness is knowing with(in) yourself. It was in the early seventeenth century that *consciousness* was coined to refer to the phenomenon of knowing-and-knowing-that-you-know. According to the *OED*, the earliest citable example of the word occurs in Philip Massinger's tragicomedy *The Maid of Honour*, published in 1632 but perhaps acted as early as 1621.[3] In that play Cariola, whose lowly rank at court gives the play its title, is courted by Bertoldo, the nobleman she in fact fancies, but she tells him she cannot entertain his suit. Why not, he asks. "The

Consciousnesse of mine owne wants" (1.2.120), she replies, alluding to the difference in their social station. Twenty years before Descartes' *cogito ergo sum*, characters in the theater were knowing-that-they-know. And they were knowing-that-they-know not only within themselves but with other actors. The fictional Cariola can appraise her "wants" only in relation to the fictional Bertoldo. The actor who played Cariola could portray that consciousness only in relation to the actor who played Bertoldo. Can you think yourself thinking? Reynolds and company say you can: "we can be aware of our consciousness, from another conscious standpoint, much the way theater makes us aware of how we script, perform, and construct our lives" (21 in this book). In that consciousness, that specifically *dramatic* consciousness, lies the possibility of liberation from the strictures of religion, law, nation, and the dogma of poststructuralist theory.

Notes

1. *Oxford English Dictionary*, con, *n.*[4], b. Abbrev. of CONFIDENCE.
2. In *Performing Transversally*, Reynolds notes that, "Guattari borrowed the term from Jean-Paul Sartre, who speaks of 'consciousness' as being composed of 'transversal' intentionalities, which are concrete and real retentions of past consciousnesses (*Transcendence of the Ego* [New York: Noonday Press, 1957]: 39)" (24), something Reynolds was not aware of when writing *Becoming Criminal*. (This endnote was added by Reynolds).
3. *Oxford English Dictionary*, consciousness, 2, with citation of Massinger's play and its printing in 1632. A first performance in 1621–22 is suggested in Philip Massinger, *Plays and Poems*, ed. Philip Edwards and Colin Gibson (Oxford: Clarendon Press, 1976), 1:xxx, 1:105.

Works cited

Ashley, Leonard R. N. *Elizabethan Popular Culture*. Bowling Green, Ohio: Bowling Green State University Press, 1988.

Massinger, Philip. *Plays and Poems*. ed. Philip Edwards and Colin Gibson. Oxford: Clarendon Press, 1976.

Reynolds, Bryan. *Becoming Criminal: Transversal Performance and Cultural Dissidence in Early Modern England*. Baltimore: Johns Hopkins University Press, 2002.

—— Reynolds, *Performing Transversally: Reimagining Shakespeare and the Critical Future*. New York: Palgrave Macmillan, 2003.

Shakespeare, William. *The Winter's Tale*. *The Complete Works*, ed. Stanley Wells and Gary Taylor. Oxford: Clarendon Press, 1988.

A Very Short List of Writings Significant to the Development of Transversal Poetics

Althusser, Louis. *Lenin and Philosophy and Other Essays*. Trans. Ben Brewster. New York: Monthly Review Press, 1971.

Anderson, Benedict. *Imagined Communities: Reflections on the Origin and Spread of Nationalism*. London: Verso, 1983.

Artaud, Antonin. *The Theater and Its Double*. Ed. and trans. Mary Caroline Richards. New York: Grove Press, 1958.

—— *Selected Writings*. Ed. Susan Sontag. Trans. Helen Weaver. Berkeley: University of California Press, 1988.

Barthes, Roland. *Mythologies*. Trans. Annette Lavers. New York: Farrar, Straus and Giroux, 1972.

Bataille, Georges. *Visions of Excess: Selected Writings, 1927–1939*. Ed. and trans. Allan Stoekl. Minneapolis: University of Minnesota Press, 1985.

Bateson, Gregory. *Steps to an Ecology of Mind*. San Francisco: Chandler Publishing, 1972.

Baudrillard, Jean. *For a Critique of the Political Economy of the Sign*. Trans. Charles Levin. St. Louis, Mo.: Telos Press, 1981.

Bergson, Henri. *Laughter: an Essay on the Meaning of the Comic*. Trans. Cloudesley Brereton and Fred Rothwell. New York: Macmillan, 1913.

Blake, William. *The Complete Poetry & Prose of William Blake*. Ed. Harold Bloom. New York: Anchor Books, 1982.

Blau, Herbert. *The Audience*. Baltimore: Johns Hopkins University Press, 1990.

Bourdieu, Pierre. *The Logic of Practice*. Trans. Richard Nice. Stanford: Stanford University Press, 1990.

—— *Language and Symbolic Power*. Ed. John B. Thompson. Cambridge: Harvard University Press, 1991.

—— *The Field of Cultural Production: Essays on Art and Literature*. Ed. and trans. Randal Jonson. New York: Columbia University Press, 1993.

Brecht, Bertolt. *Brecht on Theatre: the Development of an Aesthetic*. Ed. and trans. John Willet. London: Methuen, 1964.

Burke, Kenneth. *The Philosophy of Literary Form: Studies in Symbolic Action*. Berkeley: University of California Press, 1973.

—— *Language As Symbolic Action: Essays on Life, Literature and Method*. Berkeley: University of California Press, 1978.

Butler, Judith. *Gender Trouble: Feminism and the Subversion of Identity*. New York: Routledge, 1990.

Casey, Edward S. *The Fate of Place: a Philosophical History*. Berkeley: University of California Press, 1998.

Clément, Catherine and Hélène Cixous. *The Newly Born Woman*. Trans. Betsy Wing. London: I. B. Taurus, 1996.

Horton Cooley, Charles. *Social Organization: a Study of the Larger Mind.* Glencoe, Ill: Free Press, 1956.
—— *Human Nature and the Social Order.* New York: Schocken, 1964.
Damasio, Antonio. *Descartes' Error.* New York: Putnam, 1994.
—— *The Feeling of What Happens.* New York: Harcourt, 2000.
Davis, Angela. *Women, Race and Class.* New York: Random House, 1981.
de Certeau, Michel. *The Practice of Everyday Life.* Trans. Steven Rendall. Berkeley: University of California Press, 1984.
—— *Heterologies: Discourse on the Other.* Trans. Brian Massumi. Minneapolis: University of Minnesota Press, 1986.
—— *The Writing of History.* Trans. Tom Conley. New York: Columbia University Press, 1988.
Deleuze, Gilles. *Difference & Repetition.* Trans. Paul Patton. New York: Columbia University Press, 1994.
—— *Cinema 1: The Movement-Image.* Trans. Hugh Tomlinson and Barbara Habberjam. Minneapolis: University of Minnesota Press, 1996.
—— and Félix Guattari. *Kafka: Toward a Minor Literature.* Trans. Dana Polan. Minneapolis: University of Minnesota Press, 1986.
—— and Félix Guattari. *A Thousand Plateaus: Capitalism and Schizophrenia.* Trans. Brian Massumi. Minneapolis: University of Minnesota Press, 1987.
Dennett, Daniel. *Consciousness Explained.* Boston: Little, Brown and Company, 1991.
Derrida, Jacques. *Of Grammatology.* Trans. Gayatri Spivak. Baltimore: Johns Hopkins University Press, 1976.
—— *Writing and Difference.* Trans. Alan Bass. Chicago: University of Chicago Press, 1978.
Dewey, John. *On Education.* Ed. Reginald Archambault. Chicago: University of Chicago Press, 1964.
Dworkin, Andrea. *Pornography: Men Possessing Women.* New York: Putnam, 1981.
Dworkin, Ronald. *Taking Rights Seriously.* Cambridge: Harvard University Press, 1978.
Fiedler, Leslie A. *The Stranger in Shakespeare.* New York: Stein and Day, 1972.
Foucault, Michel. *The Archeology of Knowledge.* Trans. A. M. Sheridan Smith. New York: Pantheon Books, 1972.
—— *The Order of Things: an Archaeology of the Human Sciences.* New York: Random House, 1973.
—— *Discipline and Punish: the Birth of the Prison.* Trans. Alan Sheridan New York: Random House, 1979.
—— *The History of Sexuality: an Introduction,* Vol. 1. Trans. Robert Hurly. New York: Random House, 1990.
Glucksmann, André. *Master Thinkers.* Trans. Brian Pearce. New York: Harper & Row, 1980.
Goffman, Erving. *The Presentation of Self in Everyday Life.* New York: Anchor Books, 1959.
—— *Frame Analysis: an Essay on the Organization of Experience.* Cambridge: Harvard University Press, 1974.
Gramsci, Antonio. *Selections from Cultural Writings.* Eds. David Forgacs and Geoffrey Nowell-Smith. Cambridge: Harvard University Press, 1985.
Guattari, Félix. *Molecular Revolution: Psychiatry and Politics.* Trans. Rosemary Sheed. New York: Penguin Books, 1984.

Hegel, G. W. F. *Phenomenology of Spirit*. Trans. A. V. Miller. Oxford: Oxford University Press, 1979.

Hume, David. *Treatise of Human Nature*. Oxford: Clarendon Press, 1967.

Jameson, Fredric. *The Political Unconscious: Narrative as a Socially Symbolic Act*. Ithaca: Cornell University Press, 1981.

Kristeva, Julia. *Powers of Horror: an Essay on Abjection*. Trans. Leon S. Roudiez. New York: Columbia University Press, 1982.

Laclau, Ernesto, and Chantal Mouffe. *Hegemony and Socialistic Strategy: Towards a Radical Democratic Politics*. London and New York: Verso, 1985.

Lefebvre, Henri. *The Production of Space*. Trans. Donald Nicholson-Smith. Oxford: Blackwell, 1993.

Lyotard, Jean-François. *Libidinal Economy*. Trans. Iain Hamilton Grant. Bloomington: Indiana University Press, 1993.

MacKinnon, Catherine A. *Feminism Unmodified: Discourses on Law and Life*. Cambridge: Harvard University Press, 1987.

McLuhan, Marshall, and Quentin Fiore. *The Medium is the Massage*. New York: Bantam Books, 1967.

Herbert Mead, George. *Mind, Self and Society*. Chicago: University of Chicago Press, 1934.

Merleau-Ponty, Maurice. *Phenomenology of Perception*. Trans. Colin Smith. London: Routledge, 1996.

Nietzsche, Friedrich. *Ecce Homo*. Ed. and trans. Walter Kaufman. New York: Vintage Books, 1967.

—— *The Will to Power*. Ed. and trans. Walter Kaufmann. New York: Random House, 1968.

—— *On the Genealogy of Morals*. Ed. and trans. Walter Kaufman. New York: Vintage Books, 1989.

Nozick, Robert. *Philosophical Explanations*. Cambridge: Harvard University Press, 1981.

Poulantzas, Nicos. *Political Power and Social Classes*. New York: Verso, 1975.

Rawls, John. *A Theory of Justice*. Cambridge: Harvard University Press, 1971.

—— *Political Liberalism*. New York: Columbia University Press, 1995.

Ronell, Avital. *The Telephone Book: Technology, Schizophrenia, Electric Speech*. Lincoln: University of Nebraska Press, 1989.

Rousseau, Jean-Jacques. *The Social Contract*. Trans. by Maurice Cranston. London: Penguin, 1968.

Searle, John. *Intentionality: an Essay in the Philosophy of Mind*. Cambridge: Cambridge University Press, 1983.

—— *Minds, Brains and Science*. Cambridge: Harvard University Press, 1986.

Spinoza, Baruch. *Spinoza: Complete Works*. Ed. Michael L. Morgan, Trans. Samuel Shirley. New York: Hackett Publishing Company, 2002.

Williams, Raymond. *Marxism and Literature*. New York: Oxford University Press, 1977.

Žižek, Slavoj. *The Sublime Object of Ideology*. London and New York: Verso, 1989.

Notes on the Collaborators

Amy Cook is an Associate Instructor and doctoral student in the UCI/UCSD joint Ph.D. program in Drama and Theatre. She received her BA from the University of Michigan, Ann Arbor. She is currently working on her dissertation: "Reimagining Theater Theory: Shakespeare's Globe and the Cognitive Science Brain" (2006 expected). She has published a performance review in *Theatre Journal* and an article in *TheatreForum*. She has also directed and dramaturged in New York and San Diego (a1cook@ucsd.edu).

Donald Hedrick is Professor of English at Kansas State University and founding director of its graduate Program in Cultural Studies. He has held numerous visiting professorships in the US and abroad, including a Senior Fulbright Fellowship to Charles University in Prague. He is co-editor with Bryan Reynolds of *Shakespeare Without Class: Misappropriations of Cultural Capital* (2000), and has published essays in *PMLA*, *New Literary History*, *Shakespeare Quarterly*, *Renaissance Drama*, *English Literary History*, *English Literary Renaissance*, *Word and Image*, *College Literature*, and elsewhere on subjects from Renaissance architecture and drama to contemporary theory, film, and performance. He has directed Shakespeare in the theater, and his current research is on the origins of entertainment value (hedrick@ksu.edu).

Anthony Kubiak is Professor of Drama at the University of California, Irvine. His most recent book, *Agitated States: Performance in the American Theater of Cruelty* (2002) investigates the uses of theater and theatricality in the stagings of American history. He is also the author of *Stages of Terror: Terrorism, Ideology, and Coercion as Theatre History* (1991) and has recently published articles in the collection *Psychoanalysis and Performance* from Routledge Press, and in *Performance Research*, *TDR Theatre Journal*, *Modern Drama*, *Journal of Dramatic Theory and Criticism*, and *Comparative Drama*, among others. His current project – a series of meditations on art, nature and the Unconscious – is tentatively entitled *Theatre and Evolution: Art as Survival* (akubiak@uci.edu).

Courtney Lehmann is Associate Professor of English and Film Studies at the University of the Pacific, and she is the Director of the Pacific Humanities Center. She is an award-winning teacher and the author of *Shakespeare Remains: Theater to Film, Early Modern to Postmodern* (2002) as well as co-editor, with Lisa S. Starks, of two volumes of Shakespeare and screen criticism: *Spectacular Shakespeare: Critical Theory and Popular Cinema* (2002) and *The Reel Shakespeare: Alternative Cinema and Theory* (2002). Her most recent work, "A Thousand Shakespeares: From Cinematic Saga to Feminist Geography or, the Escape from Iceland," appears in *The Blackwell Companion to Shakespeare and Performance*, edited by Barbara Hodgdon and W. B. Worthen (2005) (clehmann@pacific.edu).

Glenn Odom is an Associate Instructor and doctoral student in Comparative Literature at the University of California at Irvine, where he is working under the supervision of Ngugi Wa Thiong'o. His forthcoming publications include *"Finding the Zumbah:* an Analysis of Infelicity in *Speech Acts in Literature"* in *Provocations to Reading: Discourses of a Democracy to Come* (2005). He completed a BA in theater, psychology, and English and a Master's degree in Secondary Education at Vanderbilt University (Odomga@hotmail.com).

Bryan Reynolds is Professor, Chancellor's Fellow, and Head of Doctoral Studies in Drama at the University of California, Irvine. He is the author of *Performing Transversally: Reimagining Shakespeare and the Critical Future* (2003), *Becoming Criminal: Transversal Performance and Cultural Dissidence in Early Modern England* (2002), co-editor with Donald Hedrick of *Shakespeare Without Class: Misappropriations of Cultural Capital* (2000), co-editor with William West of *Rematerializing Shakespeare: Authority and Representation on the Early Modern English Stage* (2005), and co-general editor with Elaine Aston of Palgrave Macmillan's new book series, *Performance Interventions*. He is also a playwright and director of theater, and a founding member of the Transversal Theater Company, whose productions have recently toured Romania and Poland (breynold@uci.edu).

Janna Segal is an Associate Instructor and doctoral student in the UCI/UCSD joint Ph.D. program in Drama and Theatre. She completed her BA in Theater at the University of California at Santa Cruz, and her MA in Theater Literature, History, Criticism, and Theory at California State University, Northridge. With Reynolds, she has published articles on *Othello* and Dario Fo's *Elisabetta* (in Reynolds, *Performing Transversally*). She was the dramaturg for the UCI and European touring productions of Reynolds' plays *Unbuckled* and *Woof, Daddy*, and is currently dramaturging the UCI production of *As You Like It*. She has also published in *Early Modern Literary Studies* (segalj@uci.edu).

Ayanna Thompson is an Assistant Professor in the Department of English at Arizona State University. Prior to coming to ASU, she taught at the University of New Mexico and Bowdoin College. She is the editor of *Shakespeare and Colorblind Casting* (2006), and she has published various articles about how issues of race are addressed through Shakespearean texts, adaptations, and appropriations. She is currently completely a book-length project, *Racing the Rack: Race and Torture on the Renaissance Stage*, which analyzes how explicit theatrical depictions of torture provide the perfect device to interrogate contradictory definitions for race (ayanna.thompson@asu.edu).

Henry S. Turner is Assistant Professor in the English Department at the University of Wisconsin-Madison, where he teaches Renaissance literature and culture and critical theory. He is the author of *The English Renaissance Stage: Geometry, Poetics and the Practical Spatial Arts* (2005) and the editor of *The Culture of Capital: Property, Cities, and Knowledge in Early Modern England* (2002). His articles have appeared or are forthcoming in *ELH, Renaissance Drama, Twentieth Century Literature,* and *The History of Cartography.* He is a contributor to the forthcoming *Norton Anthology of Drama* and is co-editor of the book series, *Literary and Scientific Cultures of Early Modernity* (Ashgate Press). He is currently writing a book on the history of the corporation in Renaissance culture (hsturner@facstaff.wisc.edu).

Additional contributor

Bruce R. Smith is College Professor of English at the University of Southern California, and the author of six books concerned with social and dramatic performance, including *Ancient Scripts and Modern Experience* (1988), *Homosexual Desire in Shakespeare's England* (1991), and *The Acoustic World of Early Modern England* (1999) (brucesmi@usc.edu).

Index

Adorno, Theodor, 230, 237, 239
aesthetic space, 155–6, 166n. 27
affective presence, x, 6, 7, 9, 20, 38,
    45–6, 85, 108n. 2, 127, 149, 153,
    244, 246, 249
All About My Mother [1999], 31
Althusser, Louis, 24n. 15, 28–9, 53nn.
    2–3, 60, 73–5, 83n. 8, 84, 129,
    139, 164n. 19, 166, 254, 257 *see
    also* Ideological State Apparatuses,
    Repressive State Apparatus
Altman, Joel, 101, 110
antitheatrical, 13, 23n. 1, 25n. 16, 62,
    82n. 1, 85, 90, 95, 108n. 3, 109n.
    13, 110–11, 158 n. 3, 167–9, 171,
    176–7, 180, 180n. 1, 181n. 5,
    182, 221n. 3, 238n. 3
antitheatricalist, 91, 101,173, 180
antitheatricality, vii, 168–9, 176–7,
    179
Aristotle, 112
Artaud, Antonin, 77, 84, 257 *see also*
    Theatre of Cruelty
articulatory space, 67, 104, 125, 127,
    136, 196, 241
As You Like It, see Shakespeare,
    William
Aunger, Robert, 4, 23nn. 4–5, 26
Austin, J. L., 30, 74–5, 249

Bacon, Sir Francis, 251
Bakhtin, Mikhail, 59n. 30, 61, 92,
    109n. 10
Balibar, Etienne, 220, 225
Bandello, Mateo, 158n. 7
Barish, Jonas, 101, 109n. 13, 110, 180
    n.1, 182
Barthelemy, Anthony Gerard, 191,
    223nn. 13–14, 225
Bartholomew Fair, see Jonson, Ben
Baston, Jane, 44–5, 58n. 28, 59n. 30,
    60
The Battle of Alcazar, see Peele, George

Baudrillard, Jean, 24n. 14, 257
Beckett, Samuel, 70
becomings, vii, 2, 5–6, 9–10, 13, 15,
    21, 45–6, 50, 66, 81, 82n. 3,
    113–14, 183, 198–200, 203,
    206–8, 219–20, 229, 231, 235,
    237, 247, 253
becomings-other, 2, 13, 66, 164n. 21,
    178, 198, 236
Bentley, Greg, 159n. 10, 163n. 18,
    164n. 20, 166
Bergson, Henri, 92, 109n. 11, 110,
    257
Berry, Herbert, 224n. 27
Bhabha, Homi, 212
Billingsley, Henry, 248
bin Laden, Osama, 38
Birringer, Johannes, 180n. 1, 181n. 3,
    182
Blackmore, Susan, 46, 15, 23n. 6, 26
Blaha, Stephen, 68, 82n. 5, 84
Blau, Herbert, 69, 82n. 7, 84, 257
Bloom, Allan, 184
Bloom, Harold, 133–4, 158n. 4,
    161–2n. 14, 167, 257
Blundeville, Thomas, 242, 248,
Boal, Augusto, 155–6, 166n. 27
Boasistuau, Pierre, 158n. 7
Bosom Buddies [Television 1980–2],
    31
Bourdieu, Pierre, 244–5, 250n. 2, 257
Boys Don't Cry [1999], 31
Brecht, Bertolt, 163n. 18, 166, 257
Brodie, Richard, 4, 23n. 5, 26
Brome, Richard, 56n. 22
Brooke, Arthur, 127, 152–3, 158–9 n.
    7, 165nn. 25–6, 166
Burt, Richard, xi, 61, 158n. 5, 166
Bush, George W., 1, 232
Butler, Judith, 11, 30, 60, 71, 73–5,
    82n. 2, 84, 111, 254, 257
Byrne, Richard W., 24n. 13, 78–9,
    83nn. 12–13, 84

Callaghan, Dympna, 129, 131–2, 157, 159n. 10, 160n. 13, 166n. 27, 238nn. 10, 12
Calvo-Merino, B., 25n. 18
carnivalesque, 92
Case, Sue-Ellen, 57
Casey, Edward, 19–20, 26, 109n. 8, 110, 112–13, 123nn. 1–2, 257
celebrity, 31–2, 42, 45–6, 242, 250nn. 1, 3
Chamberlain, John, 42, 56n. 22, 58n. 28
The Changeling, see Middleton, Thomas and Rowley, William
Cheney, Patrick, 57n. 26, 58n. 29, 60n. 33
Clark, Simon, 224n. 26
Coddon, Karen S., 238n. 10
Coleridge, Samuel Taylor, 70
comedic law, vii, 85, 89–92, 103, 107
Comensoli, Viviana, xi, 44–5, 58n. 29, 59n. 30, 60n. 33, 60
comings-to-be, x, 2, 6, 9–10, 13, 15, 21, 66, 80, 82n. 3, 114, 118, 183, 195, 199, 203–4, 206–7, 209, 219–20, 236–7, 247
conceptual territory, 11
Cook, Amy, vii, 8, 19, 85, 260
Cooke, Deryck, 227, 239
Cooper, Thomas, 88
Coulson, Seana, 92–4, 99, 103, 110
Crewe, Jonathan, 181n. 5
cross-dressing, 31, 42, 51, 58n. 28, 61, 99
The Crying Game [1992], 31
Cymbeline, see Shakespeare, William

Damasio, Antonio, 25n. 19, 26, 258
D'Amico, Jack, 195–6, 198, 222n. 11, 223n. 15, 225
Da Porto, Luigi, 159n. 7
Darwin, Charles, 23n. 8, 26, 70, 78
Davis, Lloyd, 129–31, 155, 159n. 10, 166, 258
Dawkins, Richard, 3–4, 26 *see also* meme
Day, John, 56n. 22
deceit conceits, 67, 69–70, 81, 85, 89, 91, 95, 98, 103–4, 106

deceitful imperative, 66
Deconstruction, 8–10, 252–3
Dee, John, 242, 248
De Grazia, Margreta, 46
Dekker, Thomas, 47, 96, 104,109nn. 15, 19, 110
  The Roaring Girl, vii, 27, 31–4, 38–9, 42–5, 49–51, 54n. 9, 56n. 22, 57n. 25–26, 58nn. 28–9, 59n. 30, 60n. 33, 60–2, 223n. 18, 254
  The Witch of Edmonton, 56n. 22, 108n. 5
Deleuze, Gilles, 38, 55n. 15, 62, 113, 123n. 2, 164n. 21, 198, 224n. 23, 235, 239n. 16, 258
Dennett, Daniel, 4, 23nn. 6, 8, 26n. 25, 68, 82n. 5, 84, 258
Derrida, Jacques, 9–10, 35, 41, 55n. 11, 60, 130, 254, 258 *see also différance*
De Saussure, Ferdinand, 209
De Sousa, Geraldo U., 213, 225
The Devil is an Ass, see Jonson, Ben
*différance,*10, 27, 35, 41, 252–3 *see also* Derrida, Jacques
Dionne, Craig, xi, 47, 58n. 27, 60
Dollimore, Jonathan, 44–5, 59n. 30, 60nn. 32, 33, 61, 123n. 4
Douglas, Mary, 92, 109n. 12, 110
Dressed to Kill [1980], 31
The Duchess of Malfi, see Webster, John
Dudley, Scott, 56n. 22, 233, 239

ectoplasm, 8
Eliot, T. S., 42–5, 47, 57n. 24, 61
Elizabeth I, Queen, 86, 161n. 13, 193–5, 211, 221
Elizaspace, 127
emergent activity, 21–2
emulative authority, 6, 22, 248–9
Engel, William, xi, 196, 223n. 18, 225
Enterline, Lynn, 238nn. 7,8
Everett, Barbara, 223n. 17, 225

Field, Nathaniel, 56n. 22
Fink, Bruce, 83n. 10

Fitzgerald, F. Scott, 159n. 10
Fitzpatrick, Joseph, 25n. 21, 55n.
    17, 61, 184–5, 208, 226, 239n.
    13
Ford, John
    The Witch of Edmonton, 56n. 22,
    108n. 5
Forman, Valerie, 44–5, 59nn. 29, 30,
    61
Foucault, Michel, 32, 35, 59n. 30, 61,
    110, 258
Freeman, Arthur, 191, 221nn. 5,6,
    225
Freud, Sigmund, 35, 38, 59n. 30, 61,
    66, 92, 206–8
Friar Bacon and Friar Bungay, *see*
    Greene, John viii, 240–5, 249,
    250nn 1, 3
Frith, Mary, 31–3, 39–40, 42–5, 47–9,
    56nn. 21–2, 57n. 23, 58n. 28,
    5–60n. 30, 62–3 *see also*
    Mary/Mollspace Frye, Northrop,
    127, 129, 159n. 7, 160–1nn.
    12–13, 166
fugitive elements, 7–8, 23, 86, 96,
    128, 136, 177, 184
fugitive explorations, vii, x, 1, 7–9,
    11–12, 17, 19, 22–3, 124, 128–9,
    155, 184, 208, 241
fugitive explorer, 7, 9–10, 18, 128,
    136, 253
fugitive subject, 231
fugitive subjunctivity, 17,
fugitivity, 17, 86, 93, 157, 184, 190,
    219
Fuller, David, 180n. 1, 181n. 7, 182

Galford, Ellen, 49, 61
Galilei, Galileo, 20
Gallese, Vittorio, 15, 25n. 18
Garber, Marjorie, xi, 30, 44–6, 59n.
    30, 60n. 33, 61, 114, 123, 180n.
    1, 181n. 8, 182
Garnet, Father Henry, 18
Gibbons, Brian, 159nn. 7, 9, 165n.
    24, 167
Gifford, George, 88
Glaser, D. E., 25n. 18
Gosson, Stephen, 90, 110

Greenblatt, Stephen, 110, 181nn. 1,
    4, 182, 234, 239
Greene, John, 90, 110
Greene, Robert, viii, 240–3, 245,
    248–9, 250nn. 1,3
Grezes, J., 25n. 18
Guattari, Félix, 38, 55n. 15, 62, 113,
    123n. 2, 164n. 21, 198, 224n. 23,
    235, 239n. 16, 255, 258

Habib, Imatz, 211–12, 216–17, 225
Hadfield, Andrew, 193–4, 225
Hakluyt, Richard, 195, 223n. 15
Hamlet, *see* Shakespeare, William
Hamletic guise, 65
Hamletspace, 7, 67, 127, 253
Happé, Peter, 102, 106, 108n. 4,
    109nn. 16, 18, 111
Haraway, Donna, 80–1, 84
Harriott, Thomas, 242, 248
Harris, Bernard, 195, 222n. 11, 223n.
    15
Harris, Jonathan Gil, xi, 25n. 23, 88,
    111, 181n. 9
Harsnett, Samuel, 19, 26
Hartley, Andrew James, 123n. 3
Harvey, Gabriel, 242, 248–9
Hebidge, Dick, 83n. 9
Hedrick, Donald, ii, vii, 7–8, 19, 24n.
    10, 34, 38, 55n. 13, 61, 64, 67,
    82n. 4, 84, 109n. 9, 112, 127,
    137–8, 158nn. 5–6, 166–7, 196,
    221n. 2, 224n. 21, 225n. 28, 238,
    260–1
Hegel, G. W. F., 198, 224n. 22, 226,
    239, 259
Heller, Herbert Jack, 44–5, 59n. 30,
    60–2
Henry IV, Part 1, *see* Shakespeare,
    William
Henry V, *see* Shakespeare, William
Henry VI, *see* Shakespeare, William
Heywood, Thomas, 85, 111
Holland, Norman N., 129, 161n. 14,
    167
Hopkins, D. J., 61
Hume, David, 79, 236, 239, 259
Hunter, G. K., 194, 226
hyperreal, 24n. 14

Ideological State Apparatuses, 24n. 15, 28, 83n. 8, 164n. 19, 166 *see also* Althusser, Louis
Intrilligator, James, xi, 55
investigative-expansive mode, 7, 40, 76, 86, 92, 128, 208

Jacobi, Derek, 71–2, 81, 252
Jacobs, Deborah, 44–5, 47–8, 59n. 30, 60n. 33, 61
Jaffa, Harry, 184
James I, King, 18, 99, 102, 108
Jameson, Fredric, 29–30, 32, 54n. 7, 61, 228, 238n. 4, 259
Jesus, 38
The Jew of Malta, see Marlowe, Christopher
Jones, Ann Rosalind, 109n. 16
Jones, Eldred, 223n. 19, 226
Jones, Richard, 179
Jonson, Ben, vii, 109nn. 14, 17
    Bartholomew Fair, 99
    The Devil is an Ass, vii, 8, 85–108, 108n. 4, 109n. 16, 111
    The New Inn, 101
Julius Caesar, see Shakespeare, William

Kastan, David Scott, 165n. 26, 167, 238n. 11
Kermode, Lloyd Edward, 44–6, 59n. 30, 60n. 33, 61
Kezar, Dennis, 89, 108n. 5, 111
Kidnie, Margaret Jane, 101, 109n. 16, 111
Kill Bill (dir. Quentin Tarantino), 181n. 9
Kinney, Arthur, xi, 110n. 22, 111
Knapp, Jeffrey, 192, 226
Kopel, David, 25n. 22, 111
Krantz, Susan E., 44–5, 59n. 30, 60n. 33, 62
Kristeva, Julia, 162n. 16, 163n. 17, 167, 259
Kubiak, Anthony, vii, 9, 64–5, 178, 252, 260
Kuhn, Thomas, 247
Kyd, Thomas, 189–90, 221nn. 5–7, 225
    The Spanish Tragedy, 123n. 3

Lacan, Jacques, 32, 35, 55n. 10, 62, 74, 77, 82n. 6, 83n. 10, 206–12, 224n. 25, 226
Laclau, Ernesto, 29–30, 32, 35–7, 54n. 5–6, 8, 55n. 11, 16, 62, 259
Laplanche, Jean, 224n. 25
Lehmann, Courtney, vii, 129, 159n. 10, 165n. 26, 167, 227, 260
Levin, Richard, 41, 56n. 20, 62, 213, 224n. 27, 226
Little, Arthur, 186–9, 226
Livy, 193, 198
Love's Labor's Lost, see Shakespeare, William
Lucrece, see Shakespeare, William
Lupton, Julia, 222n. 11, 223nn. 16, 19, 226
lycanthropy, 229, 235–6, 238n. 7 *see also* werewolf syndrome
Lynch, Aaron, 4, 23n. 5, 26

Macbeth, see Shakespeare, William
Macfarlane, Alan, 109n. 6, 111
Machiavellian, 9, 24n. 13, 68, 78–81, 83nn. 12, 13, 84, 146, 189, 191, 193, 198, 221nn. 5, 7, 252
Machiavellian Intelligence, 9, 24n. 13, 78–80, 83n. 12, 84
Machiavelli, Niccoli, 81, 193, 198, 222n. 10, 226
Maley, Willy, 192–3, 208, 222n. 9, 225–6
Mandeville, Sir John, 186
Marlowe, Christopher, 13, 33, 42, 51, 190
    The Jew of Malta, 185, 191
    Tamburlaine, vii, 168–80, 180–1n. 1, 181n. 2, 3, 5, 7–9, 182, 185, 222n. 14
    The Tragical History of Doctor Faustus, vii, 8, 85–7, 103–8, 110n. 20, 111, 171, 181n. 6
Marlowespace, 7, 51, 180, 253
Mary/Marlowespace, 51, 253
Mary/Mollspace, 7, 33, 39, 50 *see also* Frith, Mary
Mary, Queen of Scots, 194
Massinger, Philip, 255, 256n. 2
Master system, 229, 237

Maus, Katherine Eisaman, 43, 45, 60, 62, 123, 182, 239, 250
McKenzie, Jon, 82n. 2
McLuskie, Kathleen, 238n. 12
meme, 3–4, 23nn. 4–5, 26 *see also* Dawkins, Richard
meme machine, 5, 23n. 6, 26, 85
memeplex, 5
memetics, 4, 7
Middleton, Thomas, 47, 57n. 24, 123
  The Changeling, vii, 8, 12, 113, 116, 121–3
  The Roaring Girl, vii, 27, 31–4, 38–9, 42–5, 49–51, 54n. 9, 56n. 22, 57n. 25–6, 58nn. 28–9, 59n. 30, 60n. 33, 60–2, 223n. 18, 254
Mikalachki, Jodi, 44–5, 59n. 30, 60n. 33, 62
Mirren, Helen, 57n. 25
mirror neurons, 15, 25n. 18
Mouffe, Chantal, 29–30, 32, 35–7, 54n. 5–6, 8, 55n. 11, 16, 62, 259
Mrs. Doubtfire [1993], 31
Muenster, Sebastian, 186
Mulholland, Paul A., 44–5, 54n. 9, 56, 57n. 25, 26, 59n. 30, 62
Munday, Anthony, 108n. 3

Nakayama, Randall S., 44–5, 56n. 22, 58n. 28, 59n. 30, 62
neuro-associative conditioning, 230
The New Inn, *see* Jonson, Ben
Newton, Sir Isaac, 20–1
nodal point, 32, 35, 55n. 11
Nowell, Bradley, 124–5, 258
Nozick, Robert, 235, 239, 259

objective agency, 23n. 7, 85, 105, 108n. 1, 219, 229, 238n. 6
Odom, Glenn, vii, 8, 17, 82n. 3, 183, 243, 261
open power, 2, 206
Orgel, Stephen, 45–6, 59n. 30, 60n. 33, 62
Othello, *see* Shakespeare, William
Overbury, Sir Thomas, 98

Parsons, Robert, 192
Passingham, R. E., 25n. 18

paused consciousness, vii, 1, 13–14, 16, 21, 119, 178, 252
Peacham, Henry, ix, 213–15, 224n. 27
Peele, George, vii, 82n. 3, 183–6, 188, 190–4, 196, 199, 201–4, 208, 213, 215, 217–18, 220, 221n. 1
  The Battle of Alcazar, 185, 198, 222n. 13
performative transversations, viii, 240–1, 245, 249
place, vii, 8, 11, 14, 19–20, 26, 46, 65, 109n. 8, 110, 112–22, 123nn. 1, 2, 124, 125, 148, 153, 158n. 4, 175, 188, 190–1, 195, 211, 213, 216, 223n. 15, 227–9, 236, 241–2, 244, 248, 251–3, 257
Plutarch, 193, 198
points of resistance, 32
Pontalis, J. B., 224n. 25
Pope Pius V, 193
positive difference, 234
post-Marxist, 10, 54n. 5
post-theater, vii, 13, 168–9, 173, 176–7, 179–80
Poulantzas, Nicos, 53n. 3, 54n. 4, 62, 259
pressurized belongings, 66–7, 82n. 3, 183, 197–8, 200–1, 203–4, 206, 208–9, 211–12, 220, 243, 246, 249
principle of citationality, 245
principle of homology, 244–6
principle of translucency, 221n. 2, 225n. 28
progressive quagmire, 18
projective transversality, vii, 64, 69, 84–5, 91, 96–7, 104, 109n. 9, 221n. 2, 225n. 28

Quilligan, Maureen, 60n. 31, 60, 239
quilting point, 32, 55n. 10
Quintilian, 97–8, 111

Ramachandran, V. S., 92–4, 99, 103, 111
Rankins, William, 171, 180, 181n. 5
reflexive-consciousness, 21

rematerialization, 22, 26n. 26
Repressive State Apparatus, 24n. 15,
    28, 73, 83n. 8 *see also* Althusser,
    Louis
revenge tragedy, 123, 183, 185,
    189–91, 194, 204, 215, 221nn. 5,
    7, 223n. 15, 225
Reynolds, Bryan, ii, vii, viii, 1, 27–8,
    34–5, 38–9, 44–5, 54n. 4, 55n. 15,
    64–5, 67, 82n. 3, 84–5, 112, 124,
    127, 137–8, 168, 178, 183, 227,
    243, 250nn. 1, 3, 251–6, 260–1
  *Becoming Criminal: Transversal
    Performance and Cultural Dissidence
    in Early Modern England*, 23nn.
    1–3, 9, 24n. 11, 25nn. 15–17,
    53n. 1, 55n. 12, 59n. 30, 62,
    82n. 1, 82–3n. 8, 108nn. 1–2,
    109n. 15, 111, 143, 167, 173,
    182, 203–5, 219, 221n. 3, 223n.
    18, 226, 238nn. 3, 6, 9, 240,
    250, 256
  *Performing Transversally, Reimagining
    Shakespeare and the Critical
    Future*, 23nn. 1–3, 9, 24n. 10,
    24nn. 11–12, 25n. 15, 53n. 1,
    55n. 17, 55–6n. 18, 55n. 17, 62,
    82n. 1, 82n. 4, 84, 108nn. 1–2,
    108–9n. 6, 109n. 7, 110n. 21,
    111, 159n. 8, 167, 196, 221n. 2,
    221n. 3, 224nn. 21, 24, 226,
    238nn. 3, 9, 256
  "The Devil's House, 'or worse':
    Transversal Power and
    Antitheatrical Discourse in
    Early Modern England," 23nn.
    1–3, 24n. 11, 25n. 15, 53n. 1,
    55n. 14, 62, 82n. 1, 108n. 3,
    111, 126, 158n. 3, 164n. 21,
    167, 182, 221n. 3, 238nn. 3, 9
  with Ayanna Thompson,
    "Inspriteful Ariels: Transversal
    Tempests," 56n. 19
  with Donald Hedrick, " 'A little
    touch of Harry in the Night':
    Translucency and Projective
    Transversality in the Sexual
    and National Politics of *Henry
    V*," 84, 109n. 9, 225n. 28

  with Donald Hedrick, "Shakespace
    and Transversal Power," 24n.
    10, 55n. 13, 61, 82n. 4, 84,
    158n. 6, 167, 224n. 21, 238n.
    2
  with Donald Hedrick, *Shakespeare
    Without Class: Misappropriations
    of Cultural Capital*, 158n. 5, 166
  with D. J. Hopkins, "The Making of
    Authorships: Transversal
    Navigation in the Wake of
    Hamlet, Robert Wilson,
    Wolfgang Wiens, and
    Shakespace," 61
  with Joseph Fitzpatrick (with
    additional dialogue by Bryan
    Reynolds and Janna Segal),
    "Venetian Ideology or
    Transversal Power? Iago's
    Motives and the Means by
    which Othello Falls," 25n. 21,
    55n. 17, 61, 184–5, 208, 226,
    239n. 13
  with William West, "Shakespearean
    Emergencies: Back from
    Materialisms to Transversalisms
    and Beyond," *Rematerializing
    Shakespeare: Authority and
    Representation on the Early
    Modern English Stage*, 26n. 26
Rid, Samuel, 105, 110n. 22, 111
Riggs, David, 99, 109n. 17, 111
The Roaring Girl, see Middleton,
    Thomas, and Dekker, Thomas
R&Jspace, vii, 7, 124–5, 127–8, 136,
    138, 152, 154, 162n. 16, 165n. 26
Robbins, Anthony, 228–31, 237, 239
Roe, John, 193, 222n. 10, 226
Romeo and Juliet, see Shakespeare,
    William
Rose, Mary Beth, 43–7, 57–8n. 26,
    58n. 29, 59n. 30, 60n. 33, 62
Rousseau, Jean-Jacques, 55n. 15, 62,
    259
Rowley, William
  The Changeling, vii, 8, 12, 113, 116,
    121–3
  The Witch of Edmonton, 56n. 22, 108n.
    5

Rustici, Craig, 44–5, 59n. 30, 60n. 33, 62
Ryan, Kiernan, 129, 131–3, 162nn. 15–16, 167

Savile, Sir Henry, 248
Schluter, June, 224n. 27
Schoch, Richard, 181n. 8
Schwab, Gabrielle, 54n. 7
Searle, John, 21, 26n. 24, 259
secundum imaginationem, 90–1, 94, 99, 105, 107, 109n. 8
Segal, Janna, vii, 8, 25n. 21, 27, 55n. 17, 124, 226, 239n. 13, 254, 261
Selden, John, 99–100
Shakespace, 7, 24n. 10, 33–4, 38, 55n. 13, 56n. 19, 61, 67, 82nn. 1, 4, 84, 109n. 7, 127, 158n. 6, 162n. 16, 165n. 26, 167, 196, 224n. 21, 238, 238n. 2, 253
Shakespeare, William, 7, 23, 38, 46 58n. 27, 95, 102, 158n. 4, 228, 231, 238
  As You Like It, 31, 261
  Cymbeline, 8
  Hamlet, vii, 8–9, 20, 26, 61, 64–5, 67–72, 74 81, 82n. 6, 84, 179, 252
  Henry IV, Part 1, 67
  Henry V, 64–5, 84, 109n. 9, 221n. 2, 222n.9, 225n. 28
  Henry VI, Part 1, 20, 26
  Julius Caesar, 8
  Love's Labor's Lost, 20, 26
  Lucrece, 186–7
  Macbeth, vii, 1, 8–10, 18–19, 26, 39, 56n. 17, 72, 84–7, 103, 105–7, 109n. 6, 110–11, 180
  Othello, 17, 25n. 21, 39, 55n. 17, 61, 70, 161–2n. 14, 167, 184–5, 208, 221n. 4, 223n. 16–17, 19–20, 225–6, 239n. 13, 261
  Romeo and Juliet, vii, 8, 34, 118, 124–5, 127–38, 141–4, 148–57, 158nn. 5,6, 169nn. 7,9, 169–70n. 10, 161n. 13, 161–2n. 14, 162n. 16, 163n. 18, 164n. 20, 165n. 24, 165n. 26, 166n. 27, 167, 227

Timon of Athens, 76, 84
Titus Andronicus, vii, ix, 8, 17, 82n. 3, 133, 162n. 14, 183, 188, 214–15, 224n. 27, 226
Troilus and Cressida, 153
Twelfth Night, 31
Shakespeare In Love (dir. John Madden), 31, 137–8, 158n. 6, 162n. 16, 167, 227–8, 238n. 2
Silence of the Lambs [1991], 31
Smith, Alexander, 56n. 30
Smith, Anna Marie, 54n. 8
Smith, Bruce, viii, 251, 262
Smith, John, 102, 107–8
Smith, Paul, 32, 41, 62
Snyder, Susan, 131, 139, 141, 161–2n. 14, 164n. 20, 167
sociopolitical conductors, 1–2, 6, 12–13, 28–9, 34, 41, 44, 46, 50, 65, 67, 74, 85, 126–7, 137–9, 143, 153, 197, 213, 228, 231, 242–3, 245, 247
sociopolitical field, 24n. 15, 35, 83
somatic markers, 25n. 19
some-other-where-but-here-space, 156
Soyinka, Wole, 223n. 19
The Spanish Tragedy, see Kyd, Thomas
Spearing, Elizabeth, 44–6, 54n. 9, 56n. 22, 57n. 23, 58n. 28, 59n. 30, 63
Spinoza, Baruch, 20, 259
Stabile, Carol, 54n. 8
Stallybrass, Peter, 59n. 30, 60n. 31, 60–1, 109n. 16, 238n. 11, 239
Stamenov, Maksim, 15, 25n. 18
state machinery, 12, 24n. 15, 28, 34–5, 37, 41, 55n. 14, 67, 69, 73–4, 82n. 8, 83, 85–6, 94, 105, 108, 126, 135, 144, 147–8, 154, 158n. 3, 163n. 18, 205, 209, 228, 231–2, 237, 242, 245, 248
state power, 2, 4, 11, 16, 19, 22, 24n. 15, 28, 38–9, 53n. 3, 54n. 4, 55n. 14, 62, 75, 83n. 8, 126
Stubbes, Phillip, 108n. 3, 111, 180, 204

subjective territory, 11–12, 15–16,
21–2, 34–5, 37–42, 48, 50, 65,
67–8, 74, 76, 82n. 1, 87, 90–1, 94,
97, 105, 118, 122, 126, 135, 142,
163, 164n. 21, 178, 197, 199,
206, 208, 213, 228–30, 233,
236–7, 242–3, 255
subjunctive movement, 65
subjunctive space, vii, 16–18, 33,
40–2, 44, 48, 50, 56n. 19, 64, 65,
79, 82n. 1, 91, 94, 206, 252
subjunctivity, vii, 1, 17, 22, 41, 49,
67, 79, 90, 94, 109n. 7, 168, 207

Tamburlaine, see Marlowe,
Christopher
Taylor, Gary, 221n. 1, 256
Theater of Cruelty, 77, 84, 260 *see also*
Artaud, Antonin
theaterspace, vii, 1, 7, 13, 75, 253
Thompson, Ayanna, vii, 13, 55n. 17,
168, 186, 188, 226, 261
Titus Andronicus, see Shakespeare,
William
Todd, Janet, 44–6, 54n. 9, 56n. 22,
57n. 23, 58n. 28, 59n. 30, 63
Tokson, Elliot, 186, 188, 221n. 4, 226
Tootsie [1982], 31
Torfing, Jacob, 55n. 16,
The Tragical History of Doctor Faustus,
see Marlowe, Christopher
transcoding, 30, 32
translucent effects, 183
reading, 101
transversal agents, 33
transversal movements, 12, 14, 17,
22, 37–8, 40, 46, 67, 69, 92,
126–7
transversal poetics, vii–viii, 1, 5, 7–9,
22, 26, 28, 40, 53n. 1, 56n. 18,
63, 76, 86, 128, 159n. 8, 169,
185, 206, 208, 221n. 3, 236, 253,
257
transversal power, 2, 4, 8, 13–16, 23n.
1, 24n. 10, 25n. 21, 34, 37, 41,
55nn. 13, 17, 18, 61–2, 74–6, 81,
82nn. 1, 4, 84, 86, 108n. 3, 111,
126–7, 139, 155, 157, 158nn. 3,
6, 167, 173, 180, 182, 184, 190,

197, 206, 212, 219, 221n. 3,
224n. 21, 226, 235, 237, 238nn.
2, 3, 239n. 13, 240–1
transversal space, 38, 118, 120, 127
transversal territory, 16–18, 37, 40,
65, 82n. 1, 120, 126, 157, 169,
178, 219, 254
transversal theory, 7, 10–11, 28, 33,
37, 41, 65–7, 73, 89–91, 104, 125,
139, 159n. 8, 163n. 18, 178,
197–9, 207, 212, 228, 235, 253,
255
transvestism, 13, 31–2, 99, 252
Triumph of Love [2001], 31
Troilus and Cressida, see Shakespeare,
William
Turner, Anne, 98, 109n. 16
Turner, Henry, viii, 240, 250nn. 1, 3,
261
Twelfth Night, see Shakespeare,
William

Ungerer, Gustav, 44–5, 56n. 21, 56n.
22, 57n. 23, 58n. 28, 60n. 30,
63

Vandermast, Jacques, 244, 246–7, 249
Vickers, Brian, 41, 56n. 20, 63
Victor/ Victoria [1982], 31

Wagner, Richard, 227, 239
Webster, John, 229–231, 234, 236–8
The Duchess of Malfi, vii, 123,
222–9, 238nn. 7, 10–12, 239
The White Devil, 228, 239,
Weimann, Robert, 123n. 3,
Wells, Stanley, 129, 133–5, 158n. 4,
159n. 7, 159n. 10, 164n. 20,
166–7, 223n. 19, 225–6, 256
werewolf syndrome, 229, 236–8 *see*
also lycanthropy
West, William, ii, x, xi, 22, 26n. 26,
44–5, 59n. 30, 63, 261
Whigham, Frank, 238n. 11, 239n.
14
The White Devil, see wilderness effects,
231
Whiten, Andrew, 24n. 13, 78–9,
83nn. 12–13, 84

William Shakespeare's Romeo + Juliet
(dir. Baz Luhrman), 158n. 2, 167
Williams, Raymond, 16, 25n. 20, 40,
128–9, 159–60n. 10, 160n. 11,
167, 259
Wilson, F. P., 221n. 4
The Witch of Edmonton, see Dekker,
Thomas, Ford, John and Rowley,
William
witness-function, 177
Wittgenstein, Ludwig, 66

Wood, Ellen Meiksin, 54n. 8
Woodbridge, Linda, 55n. 12, 58n. 26,
63

Yentil [1983], 31

Zimmerman, Susan, 58n. 28, 59n. 30,
61–2
Žižek, Slavoj, 35, 63, 211, 226, 228,
238n. 5, 239, 259
zooz, 18, 26, 56n. 18, 63, 159n. 8

Printed in the United States
80395LV00001B/95